SIMULATING SCIENCE

SIMULATING SCIENCE

HEURISTICS, MENTAL MODELS, AND TECHNOSCIENTIFIC THINKING

Michael E. Gorman

INDIANA UNIVERSITY PRESS
Bloomington and Indianapolis

The paper used in this publication meets the minimum requirements of American National Standard for Information Sciences—Permanence of Paper for Printed Library Materials, ANSI Z39.48-1984.

Manufactured in the United States of America

Library of Congress Cataloging-in-Publication Data

Gorman, Michael E., date.
Simulating science : heuristics, mental models, and technoscientific thinking / Michael E. Gorman.
p. cm.—(Science, technology, and society)
Includes bibliographical references and index.
ISBN 0-253-32608-7 (alk. paper)
1. Research—Methodology. 2. Cognition. 3. Science—Philosophy. 4. Technology—Philosophy. I. Title. II. Series: Science, technology, and society (Bloomington, Ind.)
Q180.55.M4G67 1992
501—dc20 91-26630

1 2 3 4 5 96 95 94 93 92

To Margaret, who helped me begin this research,
and to Philip and Stuart,
the best products of our continuing collaboration.
Also to Bernie,
friend, colleague, and fellow-dreamer.

CONTENTS

ILLUSTRATIONS

Foreword

A foreword is a testimony, a warrant to the reader, from the perspective of great renown, that the following book is worth the time. Ideally, it should also convince the reader that there is a larger context within which the book fits, while persuading the reader of the profundity of the insights to follow, displaying, all the while, hints of even greater profundity on the part of the "foreworder." Unfortunately, I question my fitness for the task, both because I doubt my renown and because I doubt my ability to garner profound insights. Luckily for the reader, the present book should not disappoint—it *is* worth the time, and there *are* profundities in it.

I first met Mike Gorman in 1981 or 1982. I had gone to a meeting, and, after the session in which I spoke, I had been confronted by an eager young psychologist who had an uncanny familiarity with the work that Jack Mynatt, Michael Doherty, and I had conducted a few years earlier. I no longer recall the specifics of our conversation that day, but not long after, that young psychologist was in Bowling Green, joining us for Peter Wason's visit in 1982, a visit Gorman describes in chapter 2. Ever since, Mike Gorman and I have stayed in contact, visiting each other's campuses, exchanging letters and papers and (lately) E-mail, and visiting during convention and conference meetings. Mike is part of the "invisible college" of like-minded individuals who take a particular approach to the problem of understanding the psychology of science, an approach that owes its origins to Peter Wason, the distinguished British psychologist, and which spread in one of its versions to some fairly unlikely places—Bowling Green, Ohio, Houghton, Michigan, and Gainesville, Florida. The approach has now found its chronicle—the book before you. Vanity might forbid that I place Gorman as the first to take full advantage of the possible research program generated by the Bowling Green group, but that is in fact a fact. Vanity does not forbid, however, my claim that he has done far more with certain aspects of the approach than we have done: extending it, refining it, clarifying it, and linking it to questions that we never imagined were testable. Nowadays he and I enjoy more disputes over fundamental issues than we did in those not-so-long-ago "old days," and that is as it should be. But we still share a sense that there is a way to study science that relies on the methods of psychological science and that can bring a great deal of richness from the cognitive laboratory to science studies in general.

Both Gorman and I have gone on from the lab to historical studies of science, he in collaboration with Bernie Carlson and I more or less alone (though my colleagues Jack Mynatt and Mike Doherty continue the laboratory work). Gorman does not make much of the fact that his background includes training in the history of psychology at the University of New Hampshire, but it is a fact worth mentioning. The big difference between historical research and laboratory research is that in the former, contingent circumstances must be the focus of attention, whereas in the latter the goal is to procedurally filter out contingencies in the hope that universal generalizations can emerge. It has always seemed clear to me (and to Gorman) that to understand anything as complex as science, one must cover *both* the contingent and the universal. Science, on this view, is the result of universal principles acting themselves out on a stage rich with contingent circumstances. Gorman's historical training stands him in good stead in this respect; though the present book focuses upon the universal, sensitivity to the contingent is present throughout—not the least in the graceful way in which he integrates the facts of his own autobiographical circumstances with the account.

The exceptional character of the present book resides here as well; we have very few examples of books that successfully integrate autobiographical detail with scholarly treatments of a subject matter, but Gorman manages to pull it off. One aspect of his personal history is worth noting here, namely, the role of relative "marginalization" in his development. The University of New Hampshire is a fine school, and Gorman received fine training there, but it is not, with all due respect, one of the academic "Top 10." Neither is Michigan Technological University (at least not in Gorman's areas of special expertise), though it is a well-regarded engineering school. My own education was received at Wayne State University, and I have spent my entire academic career at Bowling Green State University—I too am relatively low in the "pecking order."

In conversations over the years, Gorman and I have discussed this factor and the way it may have been influencing our careers. The potentially most interesting aspect (leaving aside the obvious ones, like heavier teaching loads and lack of outside recognition) is that both he and I have enjoyed a measure of academic freedom that may have been absent in more prestigious environments. Gorman's tale of the difficulty of publishing his work (a difficulty we too experienced, probably for similar reasons) must therefore be complemented by the tale of two universities that allowed faculty members without tenure to pursue wild ideas that were very much outside the mainstream of fashionable problems at the time, universities that even rewarded such attempts. In my case at least, the intellectual ability simply would not have been there to survive and thrive in a higher-pressure environment, not, at any rate, in a muddy new area like the one I was interested in. There is a lesson here, I think, that points to potentially better policies than those that academia lives under now, especially in a day and age when niches like the ones we enjoyed are being sacrificed, even in the "non-name" universities, to the demand for ever-greater numbers of publications as a condition for tenure and promotion.

This book is therefore a normative exercise as well as a descriptive one.

Gorman makes no secret of this, of course, but the reader should not lose sight of the fact that the normative lessons extend beyond the specific issues (involving, say, falsification) that are the focus here. In addition, the book is an example of a work that, by its very existence, is a lesson.

One final point, based on a guess that I would like to hazard, a guess that may well be very wrong, that Gorman will probably disagree with, but that is interesting enough anyway to be worth considering. I believe that *Simulating Science* may well be one of the last major works in the intellectual tradition that flows from the ideas of Karl Popper. It was Popper who first articulated the value of the logic of falsification, using it to erect a criterion for the demarcation of science from non-science that has been an important touchstone for science studies ever since. Wason was among the first to see the enormous cognitive psychological implications of Popper's analysis. Wason launched a series of brilliantly simple experimental studies that put the term *confirmation bias* into the literature (actually, he used *verification bias,* a close synonym, but the other term seems to have emerged early on and has been used instead). My colleagues Mynatt and Doherty began their investigations of science by trying to investigate psychologically Kuhn's notion of a paradigm shift; if memory serves, I swayed them toward Wason's work, and it was that that made the Popperian roots of our subsequent collaborative work evident (see Mynatt et al. 1977). Since then, a number of others, sometimes independently, have seen the problem of science in Popperian terms—Wason's own student Jonathan Evans, Rich Griggs, and many others, all of whom receive due credit herein.

But Gorman may be the last, because Gorman is the one who finally establishes the limits of the view. Falsification is indeed important—psychologically, as Gorman shows, and normatively as well, but not in the simple sense in which Popper would have it. Instead, as this book well shows, falsification must take its place as *one* strategy among others, one that can sometimes entail its own biases. Confirmation must take its place as *one* among a number of possible biases, one that can sometimes entail *its* own strategic value. Gorman's work places this point in proper perspective, a broader view that suggests the real limit of Popper's view. In contrast to Popper, no single-factor account will serve to demarcate science, much less to explain its inner workings. The world is too rich and too complex for that, the world of science in particular.

Ryan D. Tweney
Bowling Green State University

Acknowledgments

Quite a few colleagues made invaluable comments on one or more chapters of this book; among them were Ryan D. Tweney, Arthur C. Houts, Ronald N. Giere, Steve Fuller, Stephen G. Brush, W. Bernard Carlson, Jeff Shrager, and Charles Bazerman. In addition to playing a crucial role in the early experiments, Margaret Gorman helped with editing and the bibliography. Thanks also to Robert Sloan and Nan Miller of the Indiana University Press for shepherding this manuscript.

Undergraduates at both Michigan Technological University and the University of Virginia played important roles in aspects of the research discussed in the book; among them were Audrey Arntzen, Jeff Boynton, Peter Carlson, Carol Christianson, Guy Cunningham, Judith Haarala, John Herman, Randy Isaacson, Ian MacKinnon, Gretchen Marz, Matthew M. Mehalik, Ann Stafford Niemiec, Michael Oblon, Virginia Patz, Mark Rathsack, James Shupe, Peggy Kingsbury Smith, Eric Stasak, Michael VeCasey, and Brent Webster. (My apologies to those whose names I have inadvertently omitted.) Working with these students was, and continues to be, a great learning experience.

Partial support for this project was supplied by the History and Philosophy of Science Program of the National Science Foundation and by the Sabbatical Leave Committee of Michigan Technological University: I am deeply grateful to both. Naturally, the views expressed herein are my own, and do not reflect the opinions of either of these organizations.

While I have not intentionally plagiarized my own writing, much of the material in this book is based on work published elsewhere and I may have inadvertently repeated a phrase here and there. The relevant sources are listed in the bibliography under my name; I would like to thank all the editors and publications for allowing me to reproduce any part of my own work.

Introduction

> Anyone proposing an empirical model of science cannot help but consider the question of how well that model fares when applied reflexively to itself. . . . In the past decade or two, the cognitive sciences have developed powerful models of human cognition which now have few competitors. This provides a strong motivation for bringing the study of science in line with these models. For science, after all, is still one of our best examples of a cognitive activity, complex though it may be.
>
> Giere 1988:384

> What I propose here, as a seventh rule of method, is in effect a moratorium on cognitive explanations of science and technology! I'd be tempted to propose a ten-year moratorium.
>
> Latour 1987:247

As the first quotation above suggests, this book will gradually build an empirical framework for studying the cognitive processes of scientists and inventors and will include a reflexive analysis of the research program out of which the framework emerged. *Reflexivity* is a loaded and somewhat dangerous term—dangerous in that it may become one of those fashionable buzzwords that everyone uses and no one understands. I am using the term *reflexive* loosely, to cover the situation where a scientist studying science studies his own experiments as examples of scientific rhetoric. The kind of reflexivity I am engaging in lies somewhere between what Woolgar (1988) calls "benign introspection" and what he calls "constitutive reflexivity." The former simply involves telling the "inside story" of how the research was done, in hopes of improving its "objectivity." The latter "amounts to a denial of distinction and a strong affirmation of similarity; representation and object are not distinct, they are intimately connected" (Woolgar 1988:22). My account is introspective and I am concerned with improving its objectivity. I also want to show how my representation of my own research was changed by writing it up.

Thus, this reflexive analysis is, in part, an answer to the moratorium proposed in the second quotation heading this introduction. Cognitive analyses generally involve distancing the observer from the observed, whereas reflexive analyses blur the observer/observed distinction. This book will do both: subject scientific reasoning to controlled laboratory study and analyze the social context out of which such studies emerge.

Therefore, this account will be both objective and personal. This book really began about fifteen years ago, when I was watching a program based on Erich Von Daniken's *Chariots of the Gods* with a group of other students at Occidental College, a school that takes pride in fostering critical thinking. The show was narrated by Rod Serling in a very dramatic voice and provided "evidence" for ancient astronauts—for example, an enormous field of uneven granite blocks which was "obviously" an alien landing field. I wondered why an advanced civilization would need to build a runway out of huge granite blocks. The show was good comedy.

After the final cut to a commercial, I was surprised to find the other students around me praising the show, saying how interesting and thought-provoking it was. Most of these students had been at Occidental for several years. Hadn't they learned how to think critically?

This incident forced me to think about what was involved in critical thinking, particularly in science. I had taken several anthropology courses. Was critical thinking primarily a function of background knowledge? But I didn't know the solutions to most of the ancient "mysteries" cited by Von Daniken—I was distrustful of the whole way in which he formulated his hypothesis and sought supporting evidence. It seemed to me that the students around me ought to be equally wary, even if they had no background in archaeology.

Some years later, in graduate school, I encountered what I thought was an instance of the same phenomenon: Walter Cronkite describing a theory that the dinosaurs had been destroyed by a meteor or comet that crashed into Earth. This seemed to me almost as bad as *Chariots of the Gods* or perhaps Immanuel Velikovsky's *Worlds In Collision:* calling on events beyond the atmosphere to explain puzzles on Earth. This hypothesis assumed that the dinosaurs had been wiped out suddenly, yet I remembered scholars arguing that their extinction was in fact gradual.

In this case, my skepticism was premature. The Alvarez hypothesis that accounts for the extinction of the dinosaurs via a radical change in Earth's climate due to a cometary collision is certainly taken seriously by scientists and is consistent with newer, more catastrophist views of evolution (Rensberger 1986), although the Alvarez hypothesis still has important rivals (Browne 1989). If I had been too critical initially, it was because I had been misled, in part, by my prior knowledge that dinosaurs had probably disappeared gradually, a position still supported by some paleontologists (Jastrow 1983).

Between these two incidents, I had begun a research program devoted to understanding the role of critical thinking in science. In the spirit of reflexivity, I thought it would be interesting to combine an account of this research with a case study of a scientist studying science: a reflexive study of a reflexive act. There is, of course, a danger in having the author of a research program describe its origins. Retrospective accounts of research are often inaccurate (see Ericsson and Simon 1984). But I have left a "documentary trail" of drafts and letters against which to check my reconstruction.

Bruno Latour, author of the second of the quotations at the beginning of this book, claims that technoscience advances in part by a strategy he calls "black-boxing," in which a device or method developed by an inventor or scientist is eventually taken for granted by others to the point where it is no longer questioned. Two examples are the double-helix structure of DNA and the Eclipse computer; after discovery and development, these two became "black boxes" for future investigators, so that a biologist can use the Eclipse to obtain "nice" pictures of DNA without having to question either the workings of the machine or the origins of the DNA theory. Similarly, the experiments reported in this book might themselves become black boxes for future investigators, so that they could simply cite Gorman's work and everyone would nod knowingly, without having to unpack how Gorman's experiments were done or what they meant at the time.

Black-boxing may be a good strategy for advancing technoscience, but it is not a good strategy for science and technology studies. Indeed, the goal of this interdisciplinary effort is to unpack as many black boxes as possible—not just the black boxes of the scientist or technologist, but also the black boxes the scholar uses to study them. Perhaps that is a good working definition of reflexivity: unpacking one's own black boxes.

Therefore, in this book I will take my own research and subject it to reflexive scrutiny, trying to prevent my experiments from being turned into black boxes. My hope is that any intelligent reader will be able to understand both the studies and the assumptions on which they were based. In addition, I will try to open the new computational black boxes being touted in science studies by Langley, Thagard, and others.

The development of my research program in many ways parallels developments in the history, philosophy, and sociology of science. My primary methodology was experimental, consistent with the tenets of logical positivism. As Houts describes it, "the positivist program assumed that the relationship between human perception and the world was virtually uncomplicated, with 'basic facts' being 'given' in 'direct observation' " (1989:53). At the time I began my experimental research, positivistic rhetoric was still very much in vogue in psychology, though it was a kind of "vulgar positivism," not closely linked to the actual philosophical positions espoused by the positivists. (For a good history of the relationship between positivism and psychology, see Smith 1986.)

I added a Popperian emphasis on the importance of disconfirmatory results and was soon embarked on a series of experiments that gradually led me to adopt views more consistent with such postpositivist philosophers as Lakatos and Kuhn—especially when I tried to generalize my results beyond the laboratory conditions under which they were obtained.

The first six chapters of this book will be devoted to a biography of this research program, covering increasingly sophisticated experimental analyses of scientific reasoning. A reflexive analysis of this research program will run concurrently with the biography, and will allow me to discuss sociological and rhetorical issues in chapter 7. Chapter 8 will include suggestions for future experimental simulations that incorporate the concerns of the new sociologists of scientific knowledge, such as Woolgar and Latour. In chapter 9 I will present and critique an alternate form of simulation: computer programs designed to model scientific discovery. Chapter 10 will move beyond the laboratory and the computer, deriving a cognitive framework from the simulations discussed in earlier chapters and applying it to a case: the invention of the telephone. This case will allow us to make some concluding remarks about the potential for cognitive studies of technoscientific creativity. Typically, issues raised in early chapters will be treated again, with greater sophistication, in later ones, mirroring the growth in my own understanding.

If this book has a moral, it concerns the value of synthesis, of trying to combine apparently incompatible perspectives—like those illustrated by the opening quotations—into a comprehensive picture. "Science scholars from different disciplines can become like the blind men studying the elephant, each one trying to describe the whole based on a study of one part. Worse, it is like adding the Tower of Babel to the metaphor of the elephant: each discipline not only develops its own perspective, but also its own language to describe it" (Gorman and Carlson 1989:89). Only through the painful process of unpacking our own and others' black boxes will we be able to determine whether we are all talking about aspects of the same technoscience and, if we are, how we can learn from each other.

SIMULATING SCIENCE

ONE

Falsification in the Laboratory

The distinction between the Chariots and Dinosaur cases corresponds to the problem of distinguishing between pseudoscience and science. This problem was especially interesting to a budding psychologist. At my alma mater, there had been a great debate over whether psychology was to be considered a science. Eventually, psychology was classed with the other social sciences—probably more for curricular than philosophical reasons. But had Occidental College chosen to consult the philosopher Karl Popper, he would have provided a justification for this decision.

1.1. POPPER'S DEMARCATION

Popper (1959; 1963) attacked the view that science advances through a steady accumulation of evidence for theories. Instead, Popper argued that a theory can never be verified, only falsified. He uses as an example a theory that all swans are white. One can encounter thousands of swans that are white and still not be sure the theory is correct, because a single encounter with a black swan would disprove it.

Therefore, Popper argues that the demarcation between scientific and nonscientific theories is that the former make predictions that may potentially be falsified, while the latter do not. As an example of a scientific theory, Popper cited General Relativity: Einstein made a specific prediction about the amount rays of light from a star would be bent by the sun's gravitational field, and the prediction was reasonably accurate. Had Einstein been wrong, General Relativity would have been disproved.

As an example of a nonscientific theory, Popper cited psychoanalysis, which has the apparent virtue of being able to explain any human action, normal or pathological. This strength is psychoanalysis's weakness: the theory forbids nothing, and therefore cannot be falsified. (But see Flanagan 1984 for evidence that psychoanalysis is falsifiable.) Popper recalls confronting Alfred Adler with a case he thought contradicted Adler's theory. After Adler had explained it, Popper asked him how he knew his explanation was correct. "Because of my thousandfold experience."

"And with this new case, I suppose, your experience has become thousand-and-one-fold," Popper responded, recognizing that every potential falsification would be transformed by Adler into a verification.

To return to the example cited in the introduction, Chariots of the Gods is a pseudoscientific theory because it forbids nothing—it makes no specific predictions about what sorts of artifacts these aliens would have left. Instead, Von Daniken takes individual pieces of evidence that he thinks cannot be explained in any other way and ties them loosely together like a "bundle of sticks" to provide evidence for his invaders. This bundle-of-sticks approach is characteristic of pseudoscience (see Kurtz 1985 for a discussion).

In contrast, one could argue that the Alvarez hypothesis does make falsifiable predictions. If a comet did radically change Earth's climate, thereby leading to the destruction of the dinosaurs, traces of the comet's destruction ought to occur in association with dramatic changes in the fossil record.

1.1.1. The Contexts of Discovery and Justification

Popper takes advantage of Reichenbach's (1938) traditional distinction between the contexts of discovery and those of justification (see Siegel 1980 for a recent defense of this distinction). According to Reichenbach, the process by which a theory is discovered need not be rational. A scientist can get a new idea from a dream, as Kekulé did when, dozing in front of the fire, he saw a snake-like chain of molecules bite its own tail and thus discovered the benzene ring (Boden 1990), or from a set of mystical or religious beliefs, as in the case of Copernicus: "The Copernican system . . . was inspired by a Neo-Platonic worship of the light of the sun who had to occupy the 'centre' because of his nobility. This indicates how myths may develop testable components. They may, in the course of discussion, become fruitful and important for science" (Popper 1963: 257).

In fact, scientists should make bold, risky conjectures, but these conjectures should have refutable consequences: "there is no more rational procedure than the method of trial and error—of conjecture and refutation; of boldly proposing theories; of trying our best to show that these are erroneous; and of accepting them tentatively if our critical efforts are unsuccessful" (Popper 1963:51). The more a theory forbids, the better it is. The process by which a discovery is evaluated or justified must be rational and depends, as we have seen, on falsification.

1.1.2. Popper and Psychology

Much of the rest of this book will be concerned with psychological studies of scientific reasoning, many of them inspired by Popper's view that falsification is the only logical way to justify a theory. Popper's attitude toward psychology should be made clear at the outset. The logic of justification is the province of philosophy. Psychology might contribute to an understanding of how scientists discover theories, but Popper was skeptical: "while the Logic of Discovery has little to learn from the Psychology of Research, the latter has much to learn from the former" (1970:58). Furthermore, he comments, "compared with physics, sociology and psychology are riddled with fashions, and with uncontrolled dogmas" (1970:57–58). As we shall see in the next section, a number of Popper's critics are more receptive to sociology and psychology.

1.1.3. A Role for Experiments

Barker (1989) points out that Popper would "reject the use of psychology in any philosophically interesting context where questions about science are to be raised" and would "deny the applicability of empirical methods to the study of science" (p. 93). The core of Popper's objection is that when one applies science to science, one is trying to understand a questionable method by using the same questionable method. How can one derive any reliable conclusions from such a study? If Popper is right, then all the experimental research cited so far is simply irrelevant to science.

However, Barker goes on to make the case that, philosophically speaking, there is no reason why scientific methods could not be used to study science. Popper's protest depends on an a priori assumption: that science can be defined as a set of rules that cut across all scientific disciplines. If so, then one must analyze these rules from a metascientific perspective such as might be provided by philosophy. But what if science is instead defined as a set of practices, or tacit knowledge? Then the knowledge and practices of psychologists can legitimately be applied to the study of science, because science is simply another form of human behavior.

Steve Fuller (1989) makes an even more radical claim: that experimental studies of scientific reasoning could serve as a basis for making normative claims about how science should be conducted. If, for example, experimental research fails to demonstrate that falsification improves problem solving on tasks that simulate scientific reasoning, one can legitimately question the normative value of Popper's prescription.

The best way to judge the potential for experimental simulations of science, given that philosophers are at least divided on their relevance, is to study them in some detail. Consequently, the rest of this chapter and the next five will involve a close look at one tradition of experimental research on scientific reasoning.

1.2. SIMULATING THE LOGIC OF JUSTIFICATION

In 1966, Wason developed a task designed to determine whether most people can falsify. In the original version of this task, subjects (people who volunteer to participate in an experiment are referred to as "subjects") are shown four cards, bearing (respectively) a vowel, a consonant, an even number, and an odd number, e.g., E, D, 4, 7. They have to select only those cards necessary to assess the rule: "If there is a vowel on one side of the card, then there must be an even number on the other."

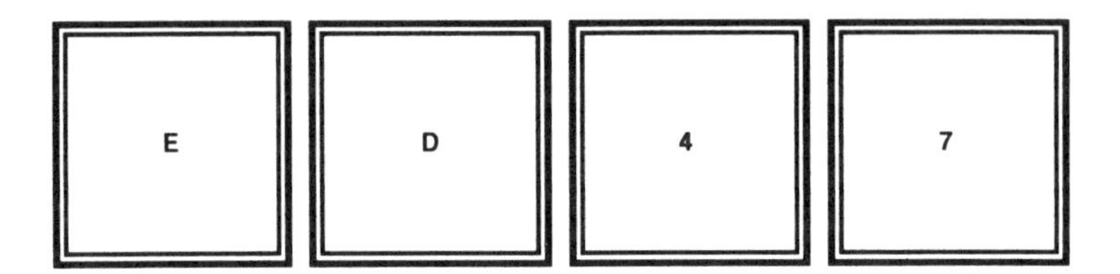

FIGURE 1-1 WASON'S SELECTION TASK
The subject is told to select only those cards necessary to assess the rule "If there is a vowel on one side, there must be an even number on the other."

This task simulates a Popperian version of the logic of justification: to justify a hypothesis, one must look at evidence that might falsify it. If there is an odd number on the other side of the vowel, then the rule has been falsified; similarly, if there is a vowel on the other side of the odd number, then the rule has been falsified. The other two cards are irrelevant.

About 50 percent of subjects pick the vowel plus the even number and another 33 percent pick the vowel card alone; only about 5 percent pick both the vowel and the odd number (Johnson-Laird and Wason 1977). Wason explained this result by supposing that subjects had a bias toward verification: they sought evidence that could confirm the hypothesis, but failed to seek evidence that could disconfirm it. The selection of the even-number card is particularly telling in this regard; it can provide additional indirect evidence for the hypothesis, but cannot disprove it. Wason (1983) reported that, on a slightly different version of this task, some subjects had trouble seeing the falsifying potential of the equivalent of the odd-numbered card even when they were allowed to look behind it and see a vowel.

Subjects do better when the task is altered in a variety of ways. For example, Griggs and Cox (1982) found that a significantly greater number of subjects in Florida solved the selection task when they were asked to test the rule "If a person is drinking beer, then the person must be over 19." The four alternatives were "drinking beer," "drinking Coke," "16 years of age," and "22 years of age." All subjects were familiar with this problem, because the drinking age in Florida was 19. As a consequence, 74 percent of the subjects solved this problem in its beer-drinking form, while none did so in its abstract vowel-and-number form. This research suggested that subjects could see the importance of the potentially falsifying instance on familiar versions of this task.[1]

1.2.1. Scientists, Statisticians, and the Selection Task

Is this task relevant to scientific reasoning? Responding to critics who "argued that the selection task lacks 'representativeness' with respect to how scientists solve problems," Wason (1983:64) said, "I would have thought that the hypothetico-deductive structure of science (according to Popper 1959) is based on the idea of falsifying the consequent of conditional sentences and so it is unclear why a miniature instantiation of such a system should be suspect." Faust (1984) makes a similar case for the relevance to science of tasks like the selection task and further argues that the cognitive abilities of scientists are not necessarily different from those of nonscientists; therefore, experiments using ordinary subjects are relevant to understanding science.

Another way of addressing the relevance issue is to study scientists directly, using these tasks. Mahoney and Kimper (1976) gave a sample of physicists, biologists, sociologists, and psychologists the selection task as part of a questionnaire, and found that only about 10 percent of the total sample—including none of the physicists—selected the two cards necessary to test the hypothesis. Kern, Mirels, and Hinshaw (1983) noted that Mahoney and Kimper's study had a low return rate; Kern et al. achieved a much higher return rate with a sample of physicists, biologists, and psychologists. Only 20 percent of the scientists selected the appropriate cards on the selection task, with physicists achieving the highest percentage (25 percent).

Einhorn and Hogarth (1978) performed a somewhat similar investigation with twenty-three statisticians. They altered the selection task to constitute an evaluation of the claim that when a particular stock market consultant says the market will rise, it always rises. The four alternatives were "favorable report" (i.e., a report by the consultant that the market will rise), "unfavorable report," "rise in the market," "fall in the market." About 20 percent of the statisticians picked the correct alternatives, i.e., the first and the last. Einhorn and Hogarth concluded that "when checking a rule concerning predictive ability, a majority of analytically sophisticated subjects failed to make the appropriate response. In particular, half of the subjects chose to examine the same piece of confirmatory information, that is, Response 1" (p. 400).

These studies suggest that scientists and statisticians may do somewhat better than ordinary subjects on the selection task: in both the Kern and the Einhorn studies, the scientists or statisticians picked the appropriate alternatives about 20 percent of the time, as opposed to a 4 percent rate for ordinary subjects in the studies summarized by Wason and Johnson-Laird (1970). While the scientists and statisticians are better than undergraduates or "the man on the street," even the professionals show signs of a confirmation bias.

Tweney and Yachanin (1985), in contrast, showed that all of a sample of ten scientists solved the selection task when it was presented in a form similar to the drinking problem posed by Griggs and Cox (1982); in a previous study, only about 25 percent of a larger sample of undergraduates had solved the same problem. A second group of ten scientists worked on a more abstract form of the selection task; 40 percent picked the two appropriate alternatives, as opposed to only about 2 percent of the undergraduates. Tweney and Yachanin's scientists were drawn from the fields of biology, chemistry, geology, and physics; all had recently published articles in refereed journals. Especially important is the fact that only 30 percent of the scientists selected the even-number card, which can confirm but not falsify, as opposed to 90 percent of the college students in the previous study.

Tweney and Yachanin also included two versions of the abstract and drinking-age problems: one in which subjects were asked to test the rule, and another in which they were asked to "find out if the rule is true or false." Each group of ten scientists experienced both wordings. There was no effect for "true or false" instructions on the drinking-age task because all scientists solved it; however, on the abstract task, the "true or false" wording increased selection of the falsifying card from 40 percent to 90 percent.

Tweney and Yachanin argue that the word *test* suggests a confirmatory heuristic to scientists: "In the usual scientific paper, a hypothesis is proposed which is 'tested' by presenting empirical results. Since there is a bias against the publication of negative results, most published articles in science present conditional inferences which are 'tested' by affirming the consequent" (1985:167). (Picking the even-number card corresponds to the fallacy of affirming the consequent.) The authors go further and suggest that this means that scientists possess the competence to perform *modus tollens*, the logic of falsification; they just do not always manifest it: "there can be no question that scientists very often ignore disconfirmatory evidence, seek evidence only to confirm their own theories, and fail to seek evidence for contrary hypotheses. The present results do not undercut this point, but they do make one thing clear: while scientists often fail to utilize logically correct rules of inference, it is *not* because they are incapable of doing so" (pp. 169–170).[2]

1.3. SIMULATING DISCOVERY AND JUSTIFICATION

While the selection task is an important tool for assessing scientists' and subjects' abilities to employ the logic of justification, it does not simulate in any way the process of rule discovery. Popper would regard this as a strength: the selection task is focused on hypothesis testing, the logical part of science. But even hypothesis testing is constrained to a set of preselected alternatives on the selection task, whereas scientists typically design their own experiments to both discover and test hypotheses. Fortunately, Wason developed another task that allows subjects and scientists to generate their own trials.

Wason developed his 2-4-6 task in 1960; it was directly inspired by a reading of Popper's (1959) *Logic of Scientific Discovery*. Subjects were told that the number triple 2,4,6 was an instance of a rule which the experimenter had in mind; they were to try to discover the rule by writing down additional number triples, with the reason they selected each triple beside it. The experimenter told them whether each triple was an instance of the rule.

For example, one subject started by writing "8,10,12" with "continuous series of even numbers" as the rationale. When the experimenter said that 8,10,12 was an instance of the rule, the subject continued to propose:

"14,16,18: even numbers in order of magnitude" (correct)

"20,22,24: same reason" (correct)

"1,3,5: two added to previous number" (correct)

At this point the subject announced that the rule was "starting with any number, two is added each time to form the next number" (Wason 1977:309). The experimenter told the subject that this guess was wrong, and the subject continued, proposing different triples and making four additional guesses, the last of which was "The rule is the three numbers must ascend in order of magnitude." This final guess was correct.

According to Wason (1977:308):

> The task was intended to simulate the understanding of an event for which several superficial explanations are possible. Since the real explanation would be merely a concealed component in the superficial ones it will most frequently defy detection until the more obvious characteristics have been varied. The analogy is not to creative thinking but to the search for simplicity, in the sense of minimal assumptions.
>
> Although the task is artificial it does possess two novel features. The correct rule cannot be proved but any incorrect hypothesis can be disproved. Moreover, an infinite number of series exemplifying any hypothesis can be generated. The subject cannot run out of numbers which confirm his hypotheses. Secondly, the subject is not shown stimuli from which he can select instances as possible evidence. He has to generate both his own hypotheses and his own evidence.

In other words, unlike the selection task, this task provides the subject with an infinite variety of possible rules and tests. Wason wondered whether subjects would settle on a rule that was sufficient to generate correct triples, or whether they would seek a rule that was both necessary and sufficient. He found that only six out of twenty-nine subjects in his original (1960) study announced the correct rule without making any incorrect guesses, and these six subjects varied their hypotheses and triples more than the others. The subject in the example above, whose first guess was that the triples went up by twos, was typical; she tried a limited range of triples similar to 2,4,6, then decided she had discovered the rule. Wason concluded that this kind of performance was evidence of a verification or confirmation bias on the part of the subjects; instead of trying to falsify or disconfirm their tentative hypotheses, they proposed triples consistent with them.

A debate sprang up almost immediately about what constituted falsification on the 2-4-6 task. The disconfirmatory alternative is clear on the selection task, but not on the 2-4-6 task: it depends on the subject's current hypothesis and the triples he or she has proposed. Wetherick (1962) and Evans (1983) argued that Wason was incorrect when he equated positive tests with confirmation and negative tests with disconfirmation. If a subject proposes "8,10,12" to test the hypothesis that the rule is "numbers go up by twos," that is a positive test—but not necessarily a confirmatory one. The subject might expect this triple to be incorrect, in which case the subject's strategy could be disconfirmatory, even though the result of the test is positive. One way to assess the subject's expectations indirectly is to look at the current hypothesis, but even if the subject's proposed triple is consistent with the current hypothesis, there is no guarantee that the subject expects this triple or hypothesis to be correct.

Furthermore, Wetherick (1962) argued, the task is biased by the initial triple. When he changed that triple to a negative instance, 4,2,6, four out of ten subjects made a correct announcement on their first guess, as opposed to only one out of ten subjects who began with 2,4,6. The sample size is too small to be significant, but the findings are suggestive, and Wason (1977) obtained a similar result when 7,5,3 was used as the starting triple. Starting with a negative instance apparently combatted subjects' tendencies to propose positive instances and make incorrect rule announcements. However, Wason argued that starting with a negative instance whose numbers were out of order merely serves to highlight the importance of order because it violates subjects' expectations—they are used to seeing numbers in order. "The rule, instead of being an implicit component in a more specific hypothesis, becomes a factor of obvious importance" (Wason 1977:312).

Furthermore, Wason (1962) pointed out that the bias in the task was deliberate; he had designed it "so that several plausible hypotheses about an unknown rule could be supported by citing instances which confirmed them, or refuted by citing instances which disconfirmed them. None of these hypotheses could be decisively confirmed, but each could be decisively eliminated by (a) citing an instance at variance with it, and (b) then observing that this instance conformed to the unknown rule." In other words, the task was designed to fit Popper's model of science: no amount of confirmation could establish the correctness of a hypothesis, but a single result at variance with the hypothesis would decisively disconfirm it.

Although the 2-4-6 task is designed to fit a Popperian framework, the fact that subjects have to generate and test their own hypotheses creates the possibility that the 2-4-6 task could be relevant to other philosophies of science. In chapter 4 we will consider Kuhn and Lakatos. But psychologists interested in science began their work on the 2-4-6 task from a naive Popperian perspective, so that is where our story properly begins.

1.3.1. Scientists and the 2-4-6 Task

Mahoney (1976) reported a study with DeMonbreun in which the performance of fifteen psychologists on the 2-4-6 task was compared with that of the same number of physicists and of Protestant ministers. Less than half of the scientists in each group eventually solved the rule, and 93 percent of the scientists returned to previously falsified hypotheses, as opposed to only 53 percent of the ministers. Mahoney concluded that scientists were more speculative and less willing to discard hypotheses than were the ministers. His (1976) account of his study is short on methodological details, but it does suggest that scientists display the same kind of confirmation bias on the 2-4-6 task as do ordinary subjects. Further research similar to the Tweney and Yachanin studies on scientists and the selection task is needed to clinch this argument.

1.3.2. Falsification in an Artificial Universe

Mynatt, Doherty, and Tweney (1977), at Bowling Green State University, explored confirmation and disconfirmation by means of an artificial universe created on a computer. Subjects fired particles at shapes on a computer screen. There were two shapes, circles and triangles, each of which could appear at one of two brightness levels, .5 or 1.0. Unbeknownst to subjects, the .5 brightness shapes were surrounded by an invisible boundary at 4.2 cm; this boundary deflected particles. All other aspects of the universe were distractors and did not affect particle motion. Mynatt et al. showed subjects two initial arrangements of shapes that biased them toward regarding triangles as deflectors of particles, because a triangle was either half lit or within the invisible boundary of a half-lit circle.

In both this artificial universe and the 2-4-6 task, subjects are given an initial instance which suggests a salient hypothesis that is not the correct rule. In the 2-4-6 task, proposing triples in accordance with this salient rule will produce an infinite number of correct instances. In Mynatt, Doherty, and Tweney's artificial universe, pursuing the salient triangle hypothesis might produce disconfirmatory results. To facilitate a comparison between confirmatory and disconfirmatory strategies, Mynatt et al. restricted subjects' choices to one of two screens arranged in pairs, ten pairs in all. One screen of each pair contained features similar to those which had stopped particles on the initial screens and hence represented

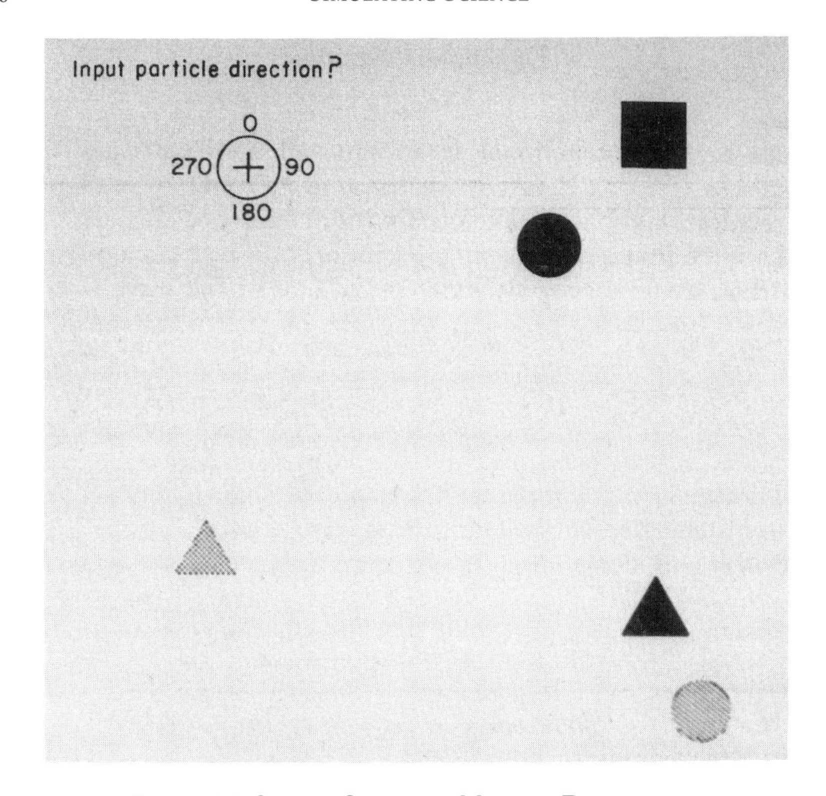

FIGURE 1-2 SECOND SCREEN IN MYNATT, DOHERTY, AND TWENEY'S ARTIFICIAL UNIVERSE
Black shapes represent objects whose brightness was 1.0; shaded figures represent 0.5 brightness. Subjects fired particles at objects from the grid on the upper left. (From Mynatt, Doherty, & Tweney 1977:89; reprinted by permission of the Experimental Psychology Society.)

a confirmatory choice; the other contained features not encountered previously, and hence represented a potentially disconfirmatory choice.

The goal of the 1977 study was to find out whether subjects preferred confirmatory or disconfirmatory information under each of three instruction conditions:

(1) Confirmatory: In this condition, "subjects were given written instructions which stated that the basic job of a scientist was to confirm theories and hypotheses" (p. 89). They were given a historical example and told to try to confirm their hypotheses about particle motion.

(2) Disconfirmatory: Instructions in this condition emphasized that scientists were supposed to disprove or disconfirm hypotheses. Again, subjects were given a historical example and told to try to follow this strategy.

(3) Control: In this condition, subjects were told that the job of a sci-

entist was to test theories and hypotheses and that they should do the same. They were given no example.

There were no significant differences across these instruction conditions. Overall, subjects selected confirmatory screens 71 percent of the time. "These results, which are remarkably similar to those found by Wason . . . suggest that confirmation bias of this sort may be a general cognitive process which is not limited to abstract tasks" (Mynatt, Doherty and Tweney 1977:93). In support of this conclusion, Mynatt et al. cited Mitroff's (1974) study of forty NASA scientists; these scientists were not only committed to confirming their own hypotheses but argued that this kind of commitment was essential to the growth of science—otherwise, good ideas would be prematurely falsified.

To summarize, Mynatt et al.'s study replicated earlier evidence of confirmation bias using a more realistic task and showed that instructions to disconfirm were insufficient to combat that bias. Those who obtained disconfirmatory information were able to make good use of it, however.

In a follow-up study, Mynatt, Doherty, and Tweney (1978) designed a more complex artificial universe and allowed subjects to interact with it under fewer constraints: instead of being forced to choose between screens, subjects could in effect design their own experiments, firing particles from any direction they chose and changing screens whenever they were ready. Particles were deflected by large, *invisible* boundaries around shapes; if a shape was at brightness 1.0, then it had no boundary. Subjects were allowed to work on the universe for a total of nine hours in separate one- to two-hour sessions and were paid for their participation.

Eight subjects, randomly selected, were given instructions modeled on Platt's (1964) "strong inference"; eight others served as a control group. Platt agreed with Popper's emphasis on disconfirmation, but added that this usually occurs in the context of alternative explanations: each experimental result can usually be explained in several ways, and future "crucial experiments" should be designed to eliminate all of these explanations but one. Platt's strong-inference heuristic involves generating competing explanations for a result, then conducting follow-up experiments that will eliminate all the explanations but one.

In Mynatt, Doherty, and Tweney's follow-up study (1978), subjects took the task very seriously; on the average, each subject performed 128 particle-firing experiments. No subject solved the system, though judges were able to rate levels of qualitative understanding. On this qualitative rating, there were no differences between strong inference and control conditions. As all subjects were taped working on the task, their protocols could be studied. In both experimental groups, more disconfirmations occurred than confirmations. This disparity was dictated in part by the universe; large, overlapping, invisible boundaries around shapes tended to falsify the most obvious hypotheses, e.g., ones based on shape.

Most subjects tended to disregard these falsifications; of 88 hypothesis

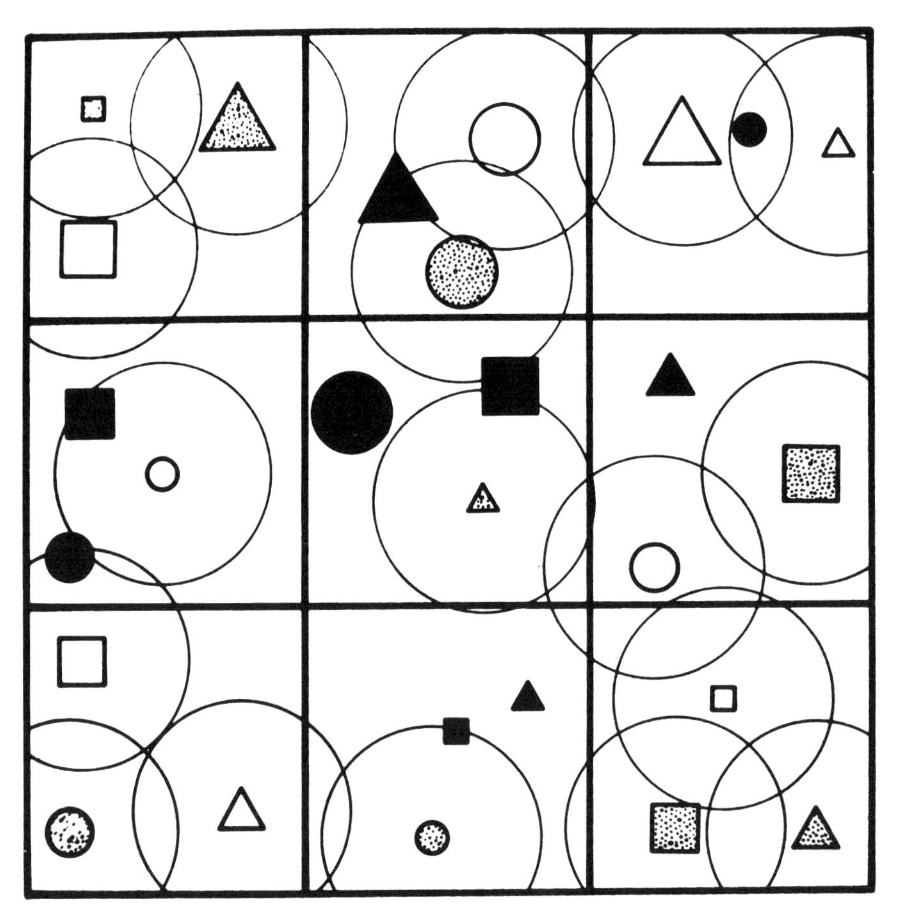

FIGURE 1-3 AN EXAMPLE OF THE NINE-SCREEN UNIVERSE USED IN TWENEY, DOHERTY, AND MYNATT
Black shapes represent 1.0 brightness, shaded 0.5, and clear .25. Boundaries, which were invisible to the subjects, have been drawn around the .5 and the .25 objects. Subjects fired particles at this computer screen; the particles were deflected by these invisible boundaries. (From Tweney, Doherty, and Mynatt 1981:149; reprinted by permission of the Columbia University Press.)

tests that resulted in disconfirmation, only 26 led to abandonment of a hypothesis. The three subjects who did abandon hypotheses immediately upon disconfirmation did not progress, however; if anything, disconfirmation led them astray by encouraging them to abandon promising hypotheses. For example, one subject decided that shapes repelled particles, but abandoned this idea after noting that in some cases particles appeared to be attracted toward shapes. These apparent instances of attraction were in fact due to the deflection of a particle off an invisible boundary. This subject's repulsion idea was correct and might have led to further promising hypotheses had he resisted disconfirmation.

Mynatt, Doherty, and Tweney concluded that confirmation may be a useful strategy early in the inference process. The most successful subject developed a hypothesis and concentrated on accumulating evidence for it; this confirmatory evidence gave him enough confidence to resist abandoning his hypothesis when disconfirmatory evidence appeared.

1.4. CONFIRMATION BIAS

The studies cited in this chapter suggested that both scientists and subjects suffered from a confirmation bias on laboratory tasks that simulate the logic of justification and also on tasks that incorporate both discovery and justification. But one could argue that as the simulations provide more room for discovery, confirmation plays an increasingly important role. In Mynatt, Doherty, and Tweney's most complicated task (1978), it was difficult to come up with a hypothesis that worked for more than a few experiments; in this kind of situation it makes sense to seek positive evidence until one has a well-corroborated hypothesis, then attempt to disconfirm it. Furthermore, even if the hypothesis is initially disconfirmed, one should resist the temptation to abandon it immediately. (A more detailed consideration of the heuristic value of confirmation will appear in 3.7.)

However, what is most striking about the Mynatt et al. studies is that instructions to disconfirm failed to improve subjects' abilities to solve the rule. Popper would probably have been indifferent to such results, but their potential implications for his philosophy are ominous: if falsification was ineffective under these ideal laboratory conditions, how could we be sure it would work in the far more noisy and complex world of actual science?

In chapter 2 I will describe some research of my own designed to determine whether subjects instructed to disconfirm perform better than those instructed to confirm. But first I need to begin the reflexive part of this book by sketching the research I was conducting at the same time as, and in complete ignorance of, the Mynatt, Doherty, and Tweney studies—research, as it happened, that looked at confirmation and disconfirmation from a different perspective.

1.5. A MAJORITY-MINORITY CONFRONTATION

My master's thesis was an effort to combine my interests in social psychology and history of science, the latter abetted by the excellent History of Psychology program at the University of New Hampshire. At the time (1977), I had not read Popper and knew nothing of the research by Wason, Tweney, Mahoney, and others. Yet I was interested in one possible solution to the Chariots problem: perhaps it was simply a question of social influence—if your friends took the idea seriously, you might, also. There was a long history of experimental literature on social influence in social psychology (see Gorman 1981 for a review of early developments), including topics like obedience (Milgram 1974), the resolution of intergroup conflict (Sherif, Harvey, White, Hood and Sherif 1961), and minority influence (Moscovici 1974).

Since I was not a Popperian, I was free to ignore the logic of justification and focus on social factors that affect the choice and defense of hypotheses. Had I but known it, some sociologists were beginning to take a similar stance at this time (see Shapin 1982 for a review), inspired in part by the philosopher Thomas Kuhn (see chapter 3).

Despite my lack of knowledge of the relevant literature, the experiment I designed contained a confirmation-disconfirmation component. The goal was to explore the relationship between confidence and influence. Asch (1956) observed that a unanimous majority could exert a strong influence on a minority of one; minority subjects publicly agreed one-third of the time with the majority's obviously incorrect judgment on a simple perceptual task. Crutchfield (1955) showed that the minority would report the correct judgment privately even as they were agreeing with the incorrect judgment publicly. Moscovici and Nemeth (1974) demonstrated that a minority which adopted a firm and consistent behavioral style could influence a majority on an ambiguous perceptual task. Mitroff (1981), in a study of lunar geologists, noted that those rated most influential by their peers were also rated most stubborn: perhaps their consistent behavioral style accounted, in part, for their influence.

None of these experiments was explicitly directed at scientific reasoning, though Crutchfield (1955) used scientists as subjects and found that they were more able to resist a majority's unanimous incorrect judgment than were other subjects. Therefore, I decided to design an experiment that simulated a majority-minority confrontation on a scientific problem. Subjects had to determine which of two theories about social structure were better predictors of whether members of a culture used consciousness-altering drugs. One theory related drug use to social structure and the other related it to religion. On each trial, subjects were presented with a case on which the theories made opposite predictions. Both the theories and the societies were constructed for the purposes of the study, though they were made as realistic as possible.

In each group, four subjects (the majority subgroup) reviewed one set of five cases, two (the minority subgroup) reviewed a different set of five, and then all six subjects met as one group to debate two more cases. Each subgroup found that all of the cases except the last one agreed with one of the two theories; if the majority found social structure was a good predictor, at the same time in another room the minority was finding that religion was a good predictor. On the last trial, the success of each subgroup was manipulated to create a two-by-two design: in one-fourth of the groups both the majority and the minority's theories were confirmed on the last trial, in one-fourth of the groups both theories were disconfirmed, in one-quarter the majority's was disconfirmed while the minority's was confirmed, and in one-quarter the minority's was disconfirmed while the majority's was confirmed.

It seemed to my thesis advisor, E. Allen Lind, and myself that the four permutations of success and failure ought to produce four patterns of influence when the majority and minority considered two final cases together:

(1) Conformity to the majority when its theory was confirmed on the final trial and the minority's disconfirmed.

(2) Minority influence when the minority obtained confirmation and the majority did not.

(3) Polarization: when both subgroups obtained confirmation, they would actually move farther apart.

(4) Normalization: when both groups' theories were disconfirmed, the groups would move closer together.

Note that a tacit assumption of this study is that confirmation increases the confidence of a scientific group or research team and disconfirmation lowers it. In other words, confirmation and disconfirmation have social effects as well as logical ones. A more radical argument would be that what constitutes a confirmation or a disconfirmation is socially negotiated, but again, I was not aware of the dawn of a new sociology of scientific knowledge that would make such claims (see chapters 7 and 8 for a discussion).

Subjects publicly voted on which theory was a better predictor for the final cases, and also indicated their preference privately on a questionnaire. Public results showed little sign of influence: majority and minority subgroups tended to support the theories they came in with. But privately, as measured on the questionnaires, the subgroups drew significantly closer by the end of the second task when the minority's theory had been falsified on the final trial (see Gorman, Lind and Williams 1977 for details of the analysis).

Group discussions should have been transcribed and analyzed to interpret these quantitative results, but crude recording equipment and time constraints prohibited. Therefore, this study remains a suggestive pilot. It may be, for example, that there are circumstances under which disconfirmation makes a minority in a scientific debate argue with greater conviction publicly, even though members have private misgivings. We will come back to some of these social influence questions in chapters 7 and 8, when we consider how experiments might be explicitly designed to complement the new sociology of scientific knowledge. Because this experiment was not specifically tailored to address such questions, it provides only indefinite answers.

The study sketched briefly above was sufficient to earn me a master's degree but I was not satisfied with it. The departure of my advisor and an inconclusive follow-up study further convinced me that this program was headed nowhere. In effect, I was disconfirming all my simple hypotheses about majority-minority interactions on these sorts of tasks, and without any confirmations it was hard for me to make sense of the results. Also, my attempts to publish my Master's results met with failure.

As Mahoney (1977) discovered, psychology journals typically are not interested in negative results. He sent a fictitious manuscript to reviewers of a behavior modification journal; each reviewer saw one of three relationships between the theoretical predictions and experimental results presented in the article: (1) results confirmed predictions; (2) results disconfirmed predictions; or (3) results were ambiguous with respect to predictions. Typically, manuscripts with positive results (condition 1 above) were recommended for acceptance, with minor revisions; manuscripts with negative or mixed results (conditions 2 and 3) were rejected. My Master's results were at least ambiguous, if not downright disconfirmations of my predictions.

Contra Mahoney, consider the results obtained in Mynatt, Doherty, and Tweney's studies (1977; 1978). In the first of these studies, instructions to disconfirm failed to improve performance; in the second, instructions to adopt a strong inference heuristic also failed. Then why were these studies accepted for publication by the *Quarterly Journal of Experimental Psychology?* In the absence of referee's reports, it is impossible to say. But one should note that Mynatt et al. framed their first study as a test of two predictions: (1) that subjects should display a confirmation bias; (2) that when they did obtain disconfirmatory information, they should be able to make appropriate use of it. Both of these predictions were confirmed. The second study did make a prediction about strong inference that was disconfirmed, but the authors were able to use task complexity to account for it. They also provided a detailed account of subjects' processes.

Therefore, disconfirmatory results can be published, if they are couched in appropriate rhetoric. One must be able to explain the disconfirmation, using subjects' processes, and establish its relevance to other researchers in the area. Like the sociology of scientific knowledge, the study of scientific rhetoric was barely in its infancy when I was struggling to publish my majority-minority research. Writing plays a major role in scientific discovery (Holmes 1987) and theory justification (Bazerman 1988). As my research program unfolds, we will explore these themes in greater detail. But my next series of experiments grew out of the Popperian perspective prevalent in the scientific reasoning literature.

TWO

Falsification and the Search for Truth

While I was still struggling with the majority-minority problem, I discovered a task that seemed to provide a better analogy to scientific reasoning than my invented societies.

2.1. THE SEARCH FOR TRUTH

In the October 1977 issue of *Scientific American,* Martin Gardner described a card game, New Eleusis, that models the scientific, mathematical, or mystical search for truth. Players in New Eleusis try to discover a rule that determines how cards can be laid out in a sequence; they take turns playing cards and the dealer ("God") tells them whether each card fits the sequence or not. For example, the dealer might decide the rule is "red and black cards must alternate." As each player puts down a card, the dealer tells them whether it fits "God's" rule or not. If it does, it is placed in sequence with the other cards that fit the rule; if it does not, it is placed at right angles to the other cards.

When I read the article I realized it would be an ideal simulation of scientific reasoning. Each card-play is a kind of experiment, and players always have a complete record of everyone's past research spread out before them.

I first attempted to set up a majority-minority confrontation with this problem, biasing one subgroup toward a color rule and the other toward a number rule, and having the two subgroups meet to debate a sequence that could fit either a color or a number interpretation. I was interested in whether the social influence processes described in section 1.5 would operate on a task explicitly designed to model the scientific search for truth.

When the subgroups came together to debate the final sequence of cards, they quickly realized that either interpretation was valid; in Popperian terms, neither subgroup's hypothesis could be falsified by the sequence of cards. Had thousands of dollars in grant money been at stake, I might have been able to engineer a lively debate. In the absence of those resources, I had to conclude that Eleusis was poorly suited to social-influence research: the evidence for or against a rule was too unambiguous.

This conclusion tallied well with traditional philosophy of science, which assumes that social explanations are necessary only when science appears irrational, e.g., when the struggle for funding forces research teams to compete even though the data suggest either interpretation is valid. However, while I was doing this research, sociologists such as Latour and Woolgar (1986) were busy demonstrating that even data can be the product of social negotiations. In chapter 4 we will introduce an important source of ambiguity into experimental simulations of scientific reasoning, and in chapters 7 and 8 we will show how experiments might help clarify the role of social negotiations in science.

But at this stage I had a task that was admirably suited to studying some aspect of the search for scientific truth, but not a majority-minority confrontation. The good Popperian scientist seems always to have a hypothesis he or she is testing; instead, I was a scientist with a simulation and no hypothesis.

Then a colleague drew my attention to Mynatt, Doherty, and Tweney (1977) and I had an idea for a new independent variable: comparing confirmatory and disconfirmatory instructions. I was also interested in a second variable that came out of the literature on group problem-solving: whether interacting groups perform better than equal numbers of individuals working separately. Perhaps groups could do what individuals had not been able to do in Mynatt et al.'s study: take advantage of instructions to disconfirm.

2.2. CONFIRMATION, COMMUNICATION, AND THE SEARCH FOR TRUTH

In my first Eleusis experiment I combined the instructional variable emphasized in Mynatt et al. with a variable that emerged out of my interest in social psychology. I compared groups of four people working together (interacting condition) to a group of four who were not permitted to communicate (co-acting condition). In interacting groups, subjects took turns playing cards, but talked openly about what they were doing and why. The experimenter gave them feedback on whether each card fit the rule or not. Interacting groups were supposed to reach consensus on the final rule on each of four Eleusis problems. Co-acting groups also took turns playing cards and saw the results obtained by others, but could not talk to one another. Each member of a co-acting group wrote his or her guess down separately.

The group conditions were crossed with two types of strategy instructions to create four combinations; co-acting confirmatory, co-acting disconfirmatory, interacting confirmatory, and interacting disconfirmatory. The goal was to see whether interacting groups would be able to falsify

Rule 1:	AD	2D	3D	4D	5S	6D	7S	6H	7H	8S	7D		
				JH	3S	8C		QC	AD				
						2H		3S					
Rule 2:	AD	2D	3D	4D	6H	8D	10H	10H	QC	JS	KD	JH	
					QC	KC		AH	8S	6S	9C		
								2D					
Rule 3:	AD	2D	3D	4D	KS	QH	7C	10S	3S	8D	AS	QH	
						10H	3C	2C	7C	QD			
							5D	6H	JD				
Rule4:	AD	2D	3D	4D	4S	10H	2C	7C	QC	10H	3H	KC	AD
				4D			8S		8C		KD		
							QC		2S		7H		

FIGURE 2-1 NEW ELEUSIS RULES

more effectively than co-acting groups. Confirmatory groups were told to test their hypotheses "by proposing as many correct cards as possible." Disconfirmatory groups were told to test their guesses by "deliberately playing cards you think will be wrong" and by studying cards that did not fit the rule. Subjects were given examples of their strategies, using "red and black cards must alternate" as a sample rule. Confirmatory groups were told that to verify this red-and-black pattern they should keep playing cards that alternated in color; disconfirmatory groups were told that the only way to be sure this pattern was the rule was to play all possible red cards after a red card and all possible black cards after a black card. Note that these instructions are consistent with Wason's (1962) use of the terms *confirmatory* and *disconfirmatory*, but that others might question whether these instructions really corresponded to what Popper meant by *confirmation* and *disconfirmation* (see section 1.1).

2.2.1. ELEUSIS: A EUREKA PROBLEM

Four Eleusis rules were used in this study; the idea was to give groups time to work together and to master their strategies. An example of each rule is shown in Figure 2-1 (where C = clubs, D = diamonds, H = hearts, S = spades, A = ace, J = jack, Q = queen, K = king). The cards in the horizontal line follow a rule; the cards in the vertical lines were incorrect when played after the card above them. Subjects were told that if the rule involved numerical values on cards, a jack would correspond to an 11, a queen to a 12, and a king to a 13.

The rules are:

1. Adjacent cards must be separated by a difference of one.
2. Adjacent cards must be separated by a difference of 0, 1, or 2.
3. Odd and even cards must alternate.
4. Cards must alternate either in terms of parity (odd vs. even) or color (red vs. black) or both.

Note that a simple sequence like "ascending diamonds" is sufficient to produce correct cards on every rule while masking the necessary rule. To find the experimenter's rule, subjects have to play some cards that should be wrong if their hypothesis is correct, e.g., proposing a KS after the 4D in Rule 2 to disconfirm the ascending diamonds hypothesis. The rules were designed to increase in complexity gradually. The first three rules emphasize differences between numbers; even Rule 3 can be rephrased as "numbers must be separated by odd differences." The fourth rule introduced an additional color dimension, but subjects had been sensitized to the possibility of a color rule by the examples that accompanied their strategy instructions.

All four rules correspond to what Steiner (1972) called a Eureka problem: one which demands a sudden insight, such as Archimedes leaping out of the bathtub with the realization that volume could be measured by displacement in water. As we will see in chapter 10, close study of the creative processes of scientists and inventors suggests that Eureka experiences are rarely as sudden or as dramatic as popular histories would have us believe, but Steiner's point was to make a contrast between Eureka and divisible tasks. The latter are particularly suited to group efforts, because each group member can tackle a portion of a larger task, whereas the former depend crucially on someone seeing a pattern. In a Eureka problem the performance of a group equals that of the best individual. On a divisible problem, which can be solved in parts, group members can be delegated subtasks and therefore will perform better than individuals working separately.

One could ask, at this point, whether scientific problems are typically Eureka or divisible. The answer is that these two categories are not entirely distinct in scientific practice. A Eureka insight on the part of one scientist depends on work done by others, and members of a research team may have insights about aspects of a complicated problem that are later combined to give a more complete view. One of the advantages of laboratory simulations is that they permit us to separate aspects of science that are hopelessly confounded with one another in actual practice; however, this artificial separation makes the process of generalizing from laboratory to life more complicated (see the beginning of chapter 4).

At any rate, it was clear to me that the Eleusis rules I was using fit Steiner's definition of a Eureka problem. Therefore, all things being equal, interacting groups ought to perform as well as the best individual in each co-acting group. But I hoped that disconfirmatory instructions would be more beneficial for groups than for individuals. After all, in science disconfirmation is frequently a group process; one theorist or research team sticks doggedly to a hypothesis while rivals seek to falsify it. In effect, disconfirmation might begin to transform a Eureka task into one that is at least partly divisible: for example, one or more group members might take on the role of falsifiers, trying to propose alternatives to the group's current hypothesis, while other members defend it vigorously.

My use of vague terms like *hoped* here is deliberate. I was not following the classic hypothetico-deductive model; instead, I had an intuition that "something interesting would happen" if I pursued this problem in this way. Popper and the other philosophers in chapter 1 talk very little about the pilot phase of research, when one is fumbling around for a promising problem or hypothesis. Laudan (1977) discusses what he calls "the context of pursuit," which is intermediate between discovery and justification. In this pursuit phase, the scientist tries to determine which hypotheses are worth the investment of time and energy involved in formal tests. I was combining two hypotheses, one about instructions and the other about type of group, in an effort to determine whether either or both were worth pursuing.

2.2.2. Subjects

Introductory psychology students in classes at the University of New Hampshire were required to participate in one or more experiments, a common practice among psychology departments. I eventually ran forty groups of four students each, ten in each cell of the two-by-two design shown below.

	Co-acting Subjects	Interacting Subjects
Confirmatory	10	10
Disconfirmatory	10	10

One might logically ask, "Why college sophomores? Why not use scientists as subjects?" Part of the answer is that it would have been impossible to ask forty scientists to spend two hours working on a laboratory simulation of scientific reasoning. Also, the purpose of the experiment was to test the effectiveness of disconfirmation under artificial conditions, not under conditions approximating real science. If disconfirmation really improved performance on tasks that simulated aspects of scientific logic, then one could argue that the improvement ought to be visible whether the subjects were scientists or not. To put it another way, scientists might already have been trained to disconfirm, in which case the strategy ought to come more naturally to them, but disconfirmatory training—if administered effectively—ought to improve the performance of intelligent college students as well.

2.2.3. Replicating Tweney and Steiner

Basically, results constituted a replication of earlier studies: disconfirmatory instructions failed to improve performance, even though disconfirmatory groups obtained significantly more incorrect cards than confirmatory groups obtained, and interacting groups performed about as well as did co-acting groups. The only hint of a difference occurred on the fourth rule, where only three confirmatory co-acting subjects solved the rule, as opposed to seven disconfirmatory co-acting subjects and an average of six out of ten interacting groups. This difference is not, however, statistically significant.[3]

Thus, Mynatt et al.'s finding that disconfirmatory instructions did not improve performance seemed robust. But the difference between the number of incorrect cards obtained by disconfirmatory and by confirmatory groups, collapsing across interacting and co-acting, was small: an average of 20.3 in disconfirmatory and 18.3 in confirmatory across all four tasks. This difference was greatest on the fourth rule—an average of 16 incorrect cards in disconfirmatory and 12.6 in confirmatory—and on this rule there was also the greatest difference in performance.

To put it another way, if number or proportion of incorrect cards is an indirect indicator of amount of disconfirmatory information, then confirmatory subjects were obtaining a fair amount; overall, 31 percent of their cards yielded disconfirmatory information, as opposed to 34 percent of the cards played by subjects actually trained to disconfirm. Did confirmatory subjects really follow their strategy?

The answer was that they had tried, but the original New Eleusis rules called for subjects to be given only thirteen cards apiece, and to receive two more cards only when they played an incorrect card. This meant that a confirmatory group that began with an "ascending diamonds" hypothesis would quickly run out of confirmatory cards and would have to play a card out of sequence. In other words, subjects instructed to confirm obtained serendipitous disconfirmations. So the apparent lack of effect for instructions might have been due to a design artifact. To test this hypothesis, I designed a follow-up study in collaboration with several colleagues (Gorman, Gorman, Latta and Cunningham 1984). But first, a word on replication.

2.2.4. Replication and Publication

Despite the possibility of an artifact, I thought the results of this initial study were worth publishing because they appeared to replicate two major findings: Mynatt, Doherty and Tweney's results regarding disconfirmatory instructions and Steiner's observations about interacting groups and individuals working separately. Psychology is a field in which major findings are often hard to replicate and, as we will see in chapter 5, replication plays a major role in eliminating potentially erroneous results.

But an attempt to publish this study as a replication of earlier research was rejected by the editor of *Social Cognition* on two grounds: (1) generally, he expected his readers would not be interested in studies of group problem-solving; (2) "your study essentially only replicates past research efforts. We already knew about the ineffectiveness of disconfirmatory strategies and the group effects you report have in principle been found before, many times" (letter from D. J. Schneider to M. E. Gorman, December 18, 1981).

Indeed, I had explicitly stressed the replicatory aspects of my experiment, citing Steiner and Tweney; I wondered to what extent I had brought the replication criticism down on myself. I had proceeded on the naive assumption that replications were worth publishing. It was one of my first encounters with the disparity between the rhetoric of science and the reality: a good Popperian would argue that replications of disconfirmations were valuable for the field: "We do not take even our own observations quite seriously, or accept them as scientific observations, until we have repeated and tested them. Only by such repetitions can we convince ourselves that we are not dealing with a mere isolated 'coincidence', but with events which, on account of their regularity and reproducibility, are in principle intersubjectively testable" (Popper 1959:45).

But my experience suggested that psychology journals were reluctant to publish replications or, in the case of *Social Cognition*, even to allow them to be refereed. Note that there were other, legitimate grounds for rejection, e.g., the thirteen-card problem noted above; it was this problem that persuaded me to publish my first Eleusis experiment only in company with a follow-up study. So from a Popperian standpoint, the journal may have been right to reject the paper, but not because it was a replication. (We will have more to say about replication and rhetoric in chapter 7.)

2.3. WHEN FALSIFICATION IS EFFECTIVE

In a follow-up study, the design of the first experiment was altered in three ways:

(1) Each subject was given a full deck of cards from which to make choices at all times; every time a card was played, it was replaced in the subject's hand. This eliminated the thirteen-card problem and made Eleusis more like the 2-4-6 task, in which a subject can always propose any experiment. Now a confirmatory group could play ascending diamonds ad nauseam.

(2) One condition or level was added to the independent variable: in addition to confirmatory and disconfirmatory instructions, one-third of the groups, randomly selected, were given instructions which urged them to try to confirm until they had a hypothesis, then disconfirm. Mynatt, Doherty, and Tweney (1978) had suggested that the most effective strategy might be this "confirm early, then disconfirm" heuristic.

(3) The co-acting condition was eliminated: all subjects were run in interacting groups. A continuation of the interacting/co-acting comparison with the thirteen-card problem eliminated might have been of interest but would have doubled the size of an already enormous experiment. I was

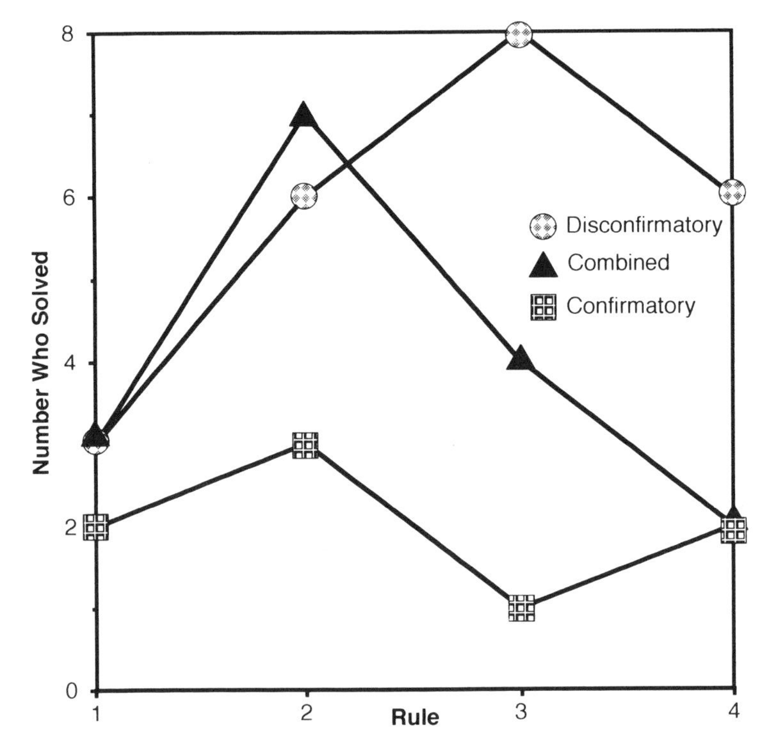

FIGURE 2-2 NUMBER OF GROUPS WHO SOLVED EACH RULE

also afraid of conducting another replication; I wanted results that could be published.

In all other respects, the second study was the same as the first. Eight groups in each strategy condition worked on the same four increasingly difficult rules. Groups were reminded of their strategy instructions between rules.

2.3.1. DISCONFIRMATION AS A SUCCESSFUL HEURISTIC

On this improved version of Eleusis, disconfirmatory instructions produced the best performance, then instructions that combined confirmation and disconfirmation, then confirmatory instructions.

Figure 2-2 shows that, as groups gained experience with their strategies, differences between conditions grew. For example, all eight disconfirmatory groups solved the third rule, "cards must alternate odd and even," as opposed to four of the combined groups and only one of the confirmatory groups. On the fourth rule, which combined color and number, disconfirmatory groups were three times as effective as groups in other conditions. These differences were statistically significant (see Gorman et al. 1984 for details), indicating that instructions to disconfirm can be very effective.

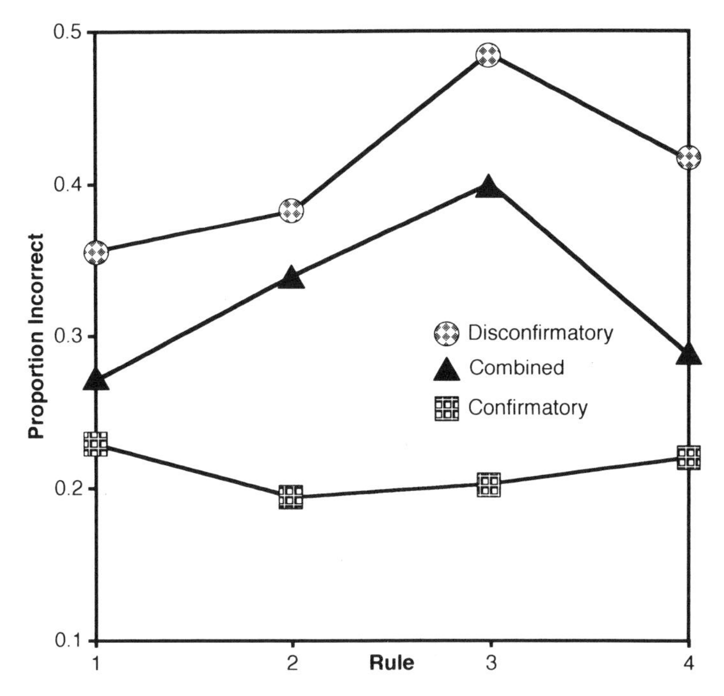

FIGURE 2-3 PROPORTION OF INCORRECT CARDS OBTAINED

The differences in solutions were mirrored by differences in proportion of incorrect cards obtained (see Figure 2-3). Proportion was used because groups in this experiment were allowed to propose as many cards as they wanted; therefore, the raw number of incorrect cards would be affected by the total number of cards. As it turned out, there were no significant differences between conditions in terms of number of cards played, but we decided the proportion measure was still more accurate and gave a better sense of the results.

Again, the disparity between disconfirmatory groups and other groups on this measure is greatest on rules three and four, where there is also the greatest disparity in performance. Note also that, unlike the previous experiment, differences between conditions are not just significant on this measure, they are dramatic: the proportion of incorrect cards obtained by disconfirmatory groups was twice that of confirmatory groups. These results indicate that both confirmatory and disconfirmatory groups followed their suggested strategies.

Consider, for example, the following cards played by one confirmatory group on Rule 3:

AD 2H 3D 4S 5C 6H 7D 8C 9C
7D

They concluded that the rule was "Cards must go in numerical order without regard to suit." Note that they played only one card that could have revealed an alternative rule—the 7D after the AD. They had followed this pattern on their first two rules and continued to confirm it on this one. Many confirmatory groups played more cards, but settled on patterns like this that were sufficient to produce correct cards but were not the necessary rule.

In contrast, consider the following cards played by a disconfirmatory group on Rule 3:

AD	2D	3D	4S	3H	2D	5D	6D	5H	6S	3H	QH	JC	10D	3D
KS		9H	4D	3C			10D					7S		JH
9H				3S								7C		5C
AH												3D		7H
												9H		JD
														9S

Like the previous group, they also began by checking for a different pattern after the AD, but when three cards failed to yield positive results, they tried the "diamonds ascending by one" pattern for two cards. Even disconfirmatory groups need some positive results to make certain that a hypothesis is worth pursuing. After the 4S, they tested for the patterns they had pursued successfully on the two previous rules. A difference of 0 worked on Rule 2, but not on this one. Cards could go down by one, as on Rules 1 and 2. After the 2D, they tried "something different": a 5D after the 2D, which was correct. When one subject several cards later suggested that "It's been even-odd, even-odd all along" another subject immediately said, "So I'll try another odd card. That should be wrong." The long chain of incorrect cards at the end indicates how persistently they tried to falsify their new hypothesis.

Group processes often facilitated falsification. Consider the following exchange in one disconfirmatory group. One member complained to the others, "I have a hard time guessing wrong." Another subject tried to help by explaining how to disconfirm: "If you think the series goes like this [pointing to a sequence of cards ascending by ones], try to prove it wrong by putting down a card that doesn't go with the series" (Gorman et al. 1984:75). The first subject took the second subject's advice, and soon both were falsifying; the group eventually solved all but the first rule.

Group processes could also interfere with falsification. In one combined group, a leader emerged who expressed Rule 2 in terms of "skips": "I think you might be able to skip cards by a certain number—like go from a four to a six but not from a four to a seven." Another subject, whom we shall designate "L," tried to restate the rule later: "Could it be the same card, or within two?" But he was ignored. L also proposed the correct solution on Rule 3, but was overruled by the leader, who kept focusing on skips and decided that an Ace (1) could follow a King (13).

By the end of Rule 3, L admitted he didn't understand the group's final guess. Here is a case in which an individual subject would have done better alone than in a group.

Overall, combined groups obtained incorrect cards on one-third of their plays, falling between confirmatory and disconfirmatory. This strategy was clearly superior to confirmation alone, but inferior to disconfirmation, perhaps because subjects' natural tendency to confirm made them attend too carefully to the confirmation part of the combined instructions. For example, one combined group began Rule 2 by proposing a long string of ascending cards, from ace to king. They knew they should be proposing more incorrect cards and at that point, one subject played a second king. When it was right, they knew that their "ascending by ones" hypothesis was wrong. They eventually solved Rules 2 and 3, but ran out of energy on Rule 4, in part because proposing long strings of confirmations takes time. This group, like most combined groups, obtained positive evidence more than two-thirds of the time. Even disconfirmatory groups sought and obtained positive evidence more than half of the time; perhaps disconfirmatory instructions are necessary if groups are going to adopt a combined strategy!

2.3.2. The Elusive Effect of Disconfirmatory Instructions

Why did disconfirmatory instructions improve performance in Gorman et al.'s (1984) Eleusis study but not in Mynatt, Doherty, and Tweney's (1977) artificial universe? There are at least three possibilities:

(1) The nature of the tasks: Perhaps the effect of disconfirmatory instructions is limited to certain tasks and situations.

(2) Groups versus individuals: Perhaps groups can falsify more effectively than individuals can.

(3) Procedural differences: Their might be other differences between Mynatt's and Gorman's studies, in terms of procedures and/or instructions.

At this point, I adopted a version of Platt's strong-inference heuristic and designed a study that I thought could potentially eliminate alternatives one and two above, which would leave three as the only possibility. I was greatly helped in this decision by Ryan D. Tweney, the third author on the Mynatt pieces; I sought him out at a convention, and he sent me reprints of some more recent work, including an attempt to use disconfirmatory instructions to improve performance on the 2-4-6 task.

2.4. DISCONFIRMATORY INSTRUCTIONS AND THE 2-4-6 TASK

Tweney, Doherty, Worner, Pliske, Mynatt, Gross, and Arkkelin (1980) conducted a series of four experiments using Wason's 2-4-6 task.

(1) The first experiment was a direct comparison of confirmatory and disconfirmatory instructions. Subjects in the confirmatory condition were given a hypothetical triple, 3,3,3, and a hypothetical rule, "three equal numbers." They were told they could test this hypothesis with triples like 8,8,8 and, if these triples were correct, they would have evidence supporting their hypothesis. Disconfirmatory subjects were given the same hypothetical triple and rule, but they were told to test it with triples like 5,7,9; if these triples were correct, then their rule would be wrong.

After completing the task, subjects were asked to go back and indicate which of their triples were intended to confirm and which to disconfirm. Instructions clearly affected strategy: disconfirmatory subjects indicated that they had intended to falsify on a significantly higher number of triples than did confirmatory subjects. Five disconfirmatory subjects solved the rule on their first guess, as opposed to two confirmatory. However, this difference was not statistically significant. Disconfirmatory instructions had a noticeable effect on strategy in this study, but not on subjects' abilities to solve the rule.

(2) Perhaps disconfirmatory instructions were being given at the wrong time. In a follow-up study, Tweney et al. withheld strategy instructions until subjects had made their first announced guess. Then subjects were given disconfirmatory instructions and told they could retract their announced rules if they liked, and continue testing. In other words, Tweney et al. gave subjects the appropriate testing heuristic at exactly the point where they were likely to need it the most. Twenty-four subjects were run under these conditions and results did not differ significantly from those in Experiment 1. Apparently, the timing of the instructions was not the problem.

(3) Tweney et al. then tried to instruct subjects to adopt Platt's strong inference as a strategy (see 1.3.2). Subjects in a strong-inference condition were told to always write down two hypotheses and two triples, one to test each hypothesis. They could discard, modify, or introduce new hypotheses, but they had to keep at least two going constantly. There was also a control condition.

Multiple-hypothesis subjects proposed significantly more triples and took significantly longer, but they solved the rule significantly less often than did control subjects. Multiple-hypothesis instructions seemed to encourage subjects to add a "dummy hypothesis" to their current hypothesis, e.g., "three even numbers in ascending order" as the real hypothesis and "three even numbers in ascending order less than 20" as the dummy hypothesis. These dummy hypotheses merely slowed subjects down and made it harder for them to discover the experimenter's rule. Of course Platt's scientists had expertise which enabled them to produce reasonable alternative hypotheses; Tweney's subjects had no particular expertise and therefore did not know what constituted a reasonable alternative. This is a point we will raise again in chapter 3, when we consider a fourth experiment by Tweney et al. (see 3.4.1).

2.4.1. When Does Disconfirmation Work?

Note that Tweney's results apparently eliminate one of the explanations for the difference between Gorman et al. (1984) and Mynatt, Doherty, and Tweney (1977); the lack of effect for disconfirmatory instructions in the latter is not due solely to a peculiarity of the artificial universe. But it leaves open the possibility that there is something about Eleusis which facilitates disconfirmation.

To find out, my wife and I (Gorman and Gorman 1984) set out to replicate either the Eleusis results or those of Tweney et al. using the 2-4-6 task. One hundred twenty subjects, drawn from a primarily engineering and science population at Michigan Technological University, were run in one of three instruction conditions: confirmatory, disconfirmatory, or control. Strategy instructions were designed to be as similar to Tweney et al.'s as possible, with the addition of a control condition in which subjects were simply told to "test" their guesses by proposing triples that would tell them whether their guess was right.

All subjects worked initially on Wason's "ascending numbers" rule, then one of two additional rules. We will consider results from the additional rules later. Results on Wason's traditional "ascending in order of magnitude" rule were dramatic: 38 out of 40 disconfirmatory subjects solved the rule, as opposed to 19 of 40 confirmatory and 21 of 40 control subjects (see Figure 2-4). Furthermore, as in Gorman et al. (1984), disconfirmatory subjects obtained incorrect triples about half the time, whereas confirmatory and control subjects were incorrect only about one-quarter of the time. Proportion of incorrect triples obtained was highly correlated with achieving the correct solution ($r = .62$).

Interestingly, control and confirmatory subjects performed about the same, which suggests that confirmation is the default heuristic on tasks of this sort.

Disconfirmatory instructions had a dramatic effect on Eleusis (see 2.3.1); this effect was replicated on the 2-4-6 task, which meant that the earlier results were not due to some peculiarity of Eleusis, or to the fact that the Eleusis study used groups and other studies used individuals.

So why were we able to obtain a significant effect for disconfirmatory instructions on the 2-4-6 task when Tweney et al. (1980) were not? Tweney's disconfirmatory subjects obtained a higher proportion of incorrect triples than did subjects in other conditions, but this additional negative information did not translate into better performance.

Initially, I thought differences in student populations might account for the differences in results. Michigan Technological University is primarily an engineering school, whereas Bowling Green State University

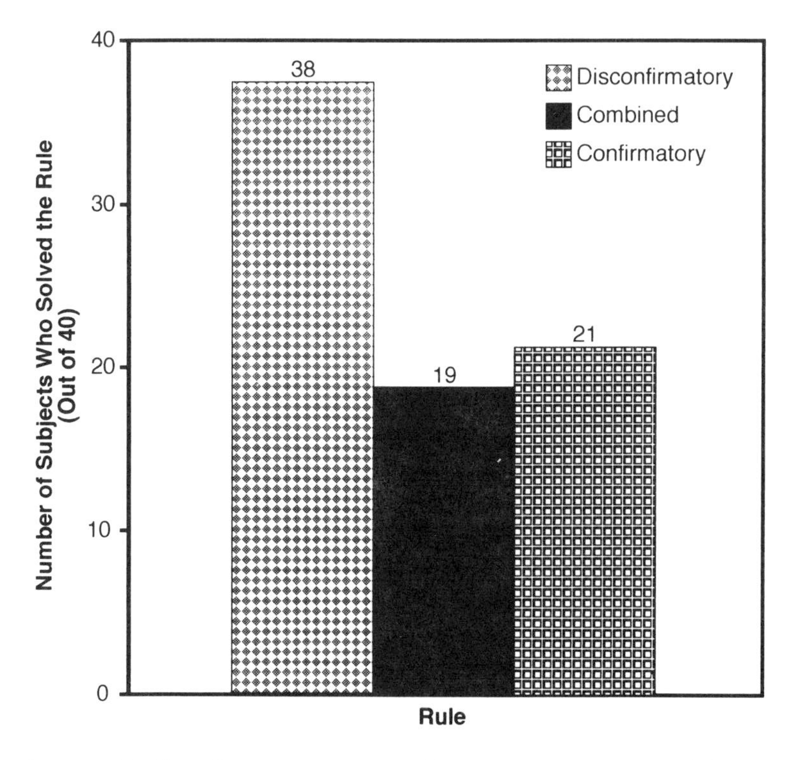

FIGURE 2-4 HOW STRATEGY AFFECTED PERFORMANCE ON THE 2-4-6 TASK

includes a more diverse range of majors. Perhaps our science and engineering students were better at falsification than were psychology students at Bowling Green. However, we had also obtained a dramatic effect for disconfirmation in our earlier Eleusis study, and this was conducted at the University of New Hampshire, a state school with a heavy population of non-science students. Once again, the close parallel in results obtained across two studies, one with groups working on Eleusis at a state school with a diverse population, the other with individuals working on the 2-4-6 task at an engineering school, suggested that neither task, nor group factors, nor student population could account for the disconfirmatory effect.

2.4.2. YOU CAN'T ASK GOD WHETHER YOUR HYPOTHESIS IS RIGHT

About the time we were getting our first results with the 2-4-6 task, I was invited to present my work to Tweney, Doherty, Mynatt, and their colleagues at Bowling Green State University in Ohio. As it happened, the dates that were best for me coincided with the time when Peter Wason would be at Bowling Green. Obviously, my work would take a back seat, but that hardly mattered compared with the honor of meeting the great man himself.

I had just submitted my first major experimental piece (eventually published as Gorman, Gorman, Latta and Cunningham 1984) to the *British Journal of Psychology;* upon arrival at Bowling Green, I found that Wason had been asked to review it. He told me he thought the work merited publication and he made some important suggestions for revision. I showed Wason and Tweney my preliminary data on disconfirmation and the 2-4-6 task; both agreed on its significance, but Wason reminded us that he wasn't really interested in whether subjects solved his rule—he was more interested in how they proceeded. He also confided that the computerized tasks developed by Mynatt, Doherty, and Tweney were not essential; his message to a young researcher was that one could do a great deal with paper-and-pencil tasks—a helpful remark, given that I was at an institution which provided little support for my experimental research.

Later, I had an opportunity to demonstrate Eleusis for Tweney. When he made a guess about the rule, I told him what I told all my subjects—that it was up to them to decide when they had found the rule. He immediately identified this as the reason I had succeeded in obtaining an effect for disconfirmatory instructions and he had failed. In Wason's traditional version of the 2-4-6 task, the experimenter always told subjects whether each of their guesses about the rule was correct or not. In effect, subjects could rely on the experimenter for falsifying information. In contrast, my subjects were given no feedback on the correctness of their hypotheses until the experiment was over, at which time all the rules were explained to them.

In actual scientific practice, the scientist has to decide whether a hypothesis is promising; he or she cannot appeal to "God" to tell her or him whether a hypothesis is right. Yet Wason and Tweney's subjects could shorten the testing process by making such an appeal. When the experimenter said the subject's hypothesis was wrong, then that subject had obtained serendipitous disconfirmation.

So it appeared that the contradiction between my experiments and Tweney's had been resolved: disconfirmatory instructions were effective when subjects received no feedback on their hypotheses from the experimenter.

2.5. INVISIBLE COLLEGES

The Bowling Green meeting illustrates the importance of what Diana Crane (1972) calls "invisible colleges"—those small groups of researchers working on similar problems who communicate regularly and are aware of each others' ongoing work long before articles appear in print. This meeting at Bowling Green brought me face-to-face with the core members of "my" invisible college.

A later experience made me aware of its boundaries. On my first 2-4-6 paper, I had to deal with the criticisms of Jonathan St. B. T. Evans, a distinguished Wason student who worked in the much larger selection-task invisible college. His major objections were epistemological, not methodological: he simply did not believe that the 2-4-6 task had very much to do with confirmation and disconfirmation (see Evans 1983; Wetherick 1962). So a reviewer outside of "my college" forced me to clarify my epistemological assumptions. Fortunately, I was able to answer his objections and the paper was subsequently published as Gorman and Gorman 1984.[4]

An invisible college is often started by an individual like Wason, whose classic papers form the basis of both the 2-4-6 and the selection-task colleges. Students of the founder, like Evans, and independent colleagues, like the Bowling Green group, play an important role in propagating the college, whose membership on the periphery remains fluid but whose core members are easy to identify. From a Popperian standpoint, this whole concept of an invisible college is hopelessly vague and illustrates the weakness of sociological and psychological approaches. Yet metaphorically the term makes sense: there are clusters of scholars who communicate informally, cite the same body of papers, and identify themselves as working on the same problem.[5] The reception of my Popperian research was forcing me to consider the ways in which the social and cognitive aspects of science are intertwined (see chapter 7).

2.6. FALSIFICATION IN SCIENCE AND SIMULATION

Research in this chapter has suggested that, in situations where one cannot ask "God" whether one's hypothesis is correct, disconfirmation is a successful heuristic. So far, experimental simulations seemed to support a Popperian view of science. But my own publication experiences suggested that scientific practice did not correspond with Popper's prescriptions: it was difficult to publish replications and falsifications, and reviewers' objections had more to do with their invisible colleges than with any logic of falsification. Popper might be perturbed by the less than ideal performance of scientists, but he would simply stress that falsification was intended as a normative ideal, not as a description of actual scientific practices. As we shall see in the next chapter, Popper's critics were questioning even the normative value of falsification.

THREE

When Falsification Fails

When I designed the 2-4-6 study reported in the last chapter, I decided to have subjects work on a second rule, in addition to the traditional "ascending numbers." In the Eleusis experiment, the effect of instructions had not appeared until later rules. I wondered if the same pattern would occur on the 2-4-6 task. It didn't. In fact, the pattern was reversed—disconfirmatory instructions made a great difference on the first rule, but no difference on the second one. This chapter begins with a description of these results, then describes how this paradox motivated a series of experiments that forced me to clarify the relationship between disconfirmation and mental representations.

3.1. TWO SECOND RULES

In addition to the initial "ascending numbers" rule, which was encountered by all 120 subjects in Gorman and Gorman 1984, I used two different second rules. Forty-eight of these 120 subjects encountered "any triple with at least one even number in it will be correct" as their second rule. I selected this rule somewhat arbitrarily; at that point I was mainly concerned with whether they solved the first rule. (Note that subjects were not told whether they had solved the first rule before starting on the second.) Ten confirmatory, eleven disconfirmatory, and seven control subjects (out of sixteen in each condition) solved the "at least one even number" rule; this difference was not statistically significant. There was no significant difference in proportion of incorrect triples obtained either, despite the fact that on the first rule there were strong differences on both of these measures.

Perhaps falsification failed because a confirmatory heuristic would provide subjects with disconfirmatory information on the "at least one even number" rule. A subject who decided to try the same hypothesis as he or she used in the first task might begin by proposing all ascending numbers; as soon as he or she proposed three odd numbers, e.g., 1,3,5, the ascending pattern would be disconfirmed. Support for this notion comes from the fact that the proportion of incorrect triples obtained was not significantly correlated with success on this rule. Subjects could "stumble on" the correct rule without obtaining a large proportion of instances at variance with their hypotheses.

Therefore, I decided to use a different second rule for the remaining subjects in the "ascending numbers" study. Furthermore, I realized it was time to think more carefully about the relationship between the first and second rules. I decided to make the second rule sufficiently general so that every instance that was correct on Wason's original rule would also be correct on this one. Disconfirmation was essential on "ascending numbers" because subjects usually began with hypotheses that were subsets of that rule; therefore, these hypotheses would produce an infinite number of correct instances. I wanted a second rule that would force even those subjects who solved the first to propose disconfirmatory triples if they wanted to solve the rule. This meant that the second rule had to be so general that "ascending numbers" would be a subset of it.

So I decided to try "three numbers must be different" as a rule. If a subject followed an "ascending numbers" pattern on this rule, he or she would be able to obtain an infinite number of correct triples; incorrect triples would have to have two or three numbers the same. This rule was solved by only ten of twenty-four confirmatory subjects, ten of twenty-four control subjects, and fourteen of twenty-four disconfirmatory subjects. This difference was not significant (see Figure 3-1).

Similarly, 17 percent of the triples proposed by confirmatory and 17 percent by control subjects were incorrect, as opposed to 24 percent proposed by disconfirmatory subjects; again, this difference was not statistically significant. But proportion of incorrect triples was highly correlated with success on this rule ($r = .73$). Had disconfirmatory subjects successfully applied their strategy on this rule, it would have improved their performance.

As a strategy check, we had used another measure in this study. We asked subjects to predict whether each triple they proposed would be correct or incorrect. Differences on this measure were highly significant on the "ascending numbers" rule and on both second rules (see Gorman and Gorman 1984 for details); on the "three different numbers" rule, disconfirmatory subjects predicted that 59 percent of their triples would be wrong, as opposed to 29 percent for confirmatory and 35 percent for control subjects. Disconfirmatory subjects were obviously trying to propose incorrect triples. The problem was, they couldn't obtain them (see Figure 3-2).

Thus, the strong effect for disconfirmatory instructions on Wason's "ascending numbers" rule disappeared on later rules. In contrast, disconfirmatory groups working on Eleusis rules (Gorman et al. 1984) became more and more proficient at using their strategy as they progressed to more difficult rules. Was this difference between the studies due to the fact that groups, as opposed to individuals, get better with practice? Was

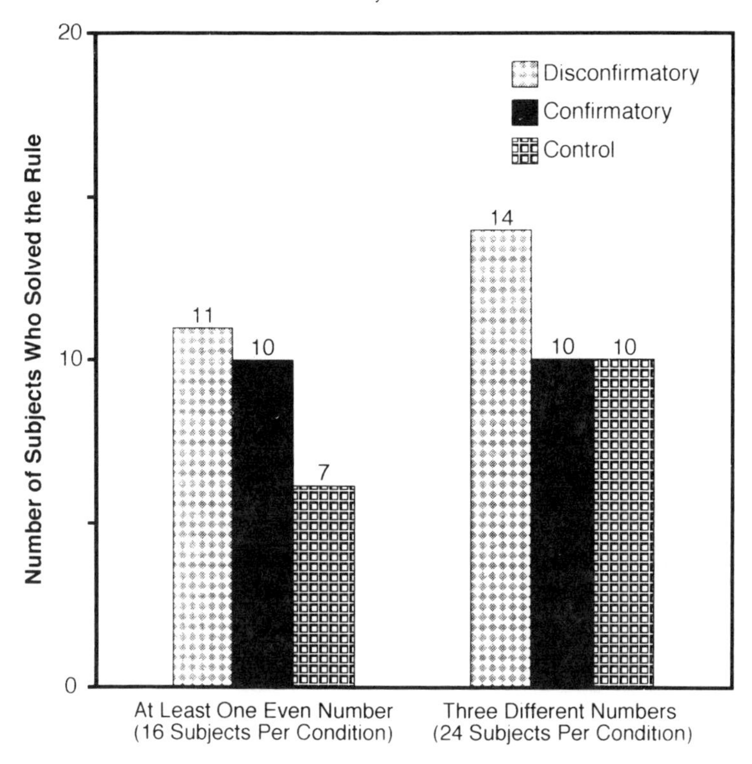

FIGURE 3-1 HOW STRATEGY AFFECTED PERFORMANCE ON TWO SECOND RULES

it some feature of the two tasks and/or the specific rules used on each task? These issues could not be settled on the basis of these two studies; future research would be required.

In particular, we wondered if there was an order effect: perhaps working on the "ascending numbers" rule gave subjects a representation of the problem that made it particularly hard for them to solve the "three different numbers" rule. The four Eleusis rules were designed to focus subjects on differences between adjacent cards, which facilitated their performance on later rules. In contrast, the "ascending numbers" rule might have focused subjects on order, a dimension that is irrelevant on the "three different numbers" rule.

To find out, we decided to conduct a follow-up study in which the order of these two rules would be reversed, i.e., the "three different numbers" rule would come before the "ascending numbers" rule. If disconfirmatory instructions were effective across a wide range of tasks, then we would expect them to produce superior performance on the "three different numbers" rule when it was uncontaminated by any previous rule.

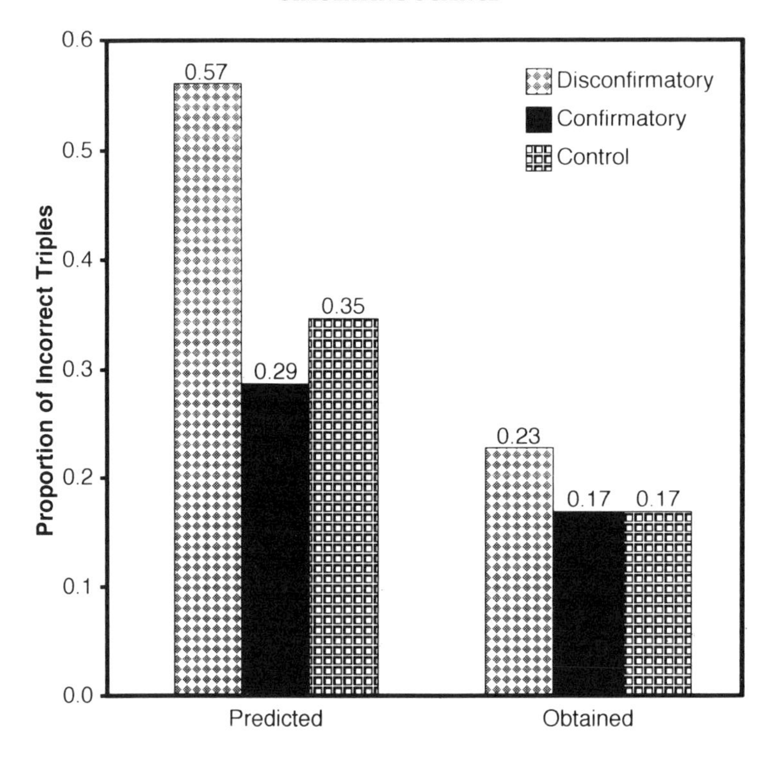

FIGURE 3-2 A COMPARISON OF PREDICTED AND OBTAINED INCORRECT TRIPLES ON THE "THREE DIFFERENT NUMBERS" RULE

3.2. DOES DISCONFIRMATION DEPEND ON ORDER?

When the order of the rules was reversed, not only were there still no differences between strategy conditions in terms of solutions on the "three different numbers" rule, the experiment with this first rule wiped out the effect of disconfirmatory instructions on the "ascending numbers" rule (see Figure 3-3).

Similarly, there were no significant differences between conditions in terms of number of incorrect triples obtained: on the "three different numbers" rule, subjects in all three conditions obtained about the same proportion of incorrect triples, one-fifth; on the "ascending numbers" rule, disconfirmatory and control obtained incorrect triples about half of the time and confirmatory were incorrect on about one-third of their trials (see Gorman, Stafford, and Gorman 1987 for details). Disconfirmatory subjects did predict that significantly more of their triples would be incorrect on both rules, indicating that they intended to disconfirm.

Overall, results on the "three different numbers" rule were similar across both studies: regardless of the order in which they encountered

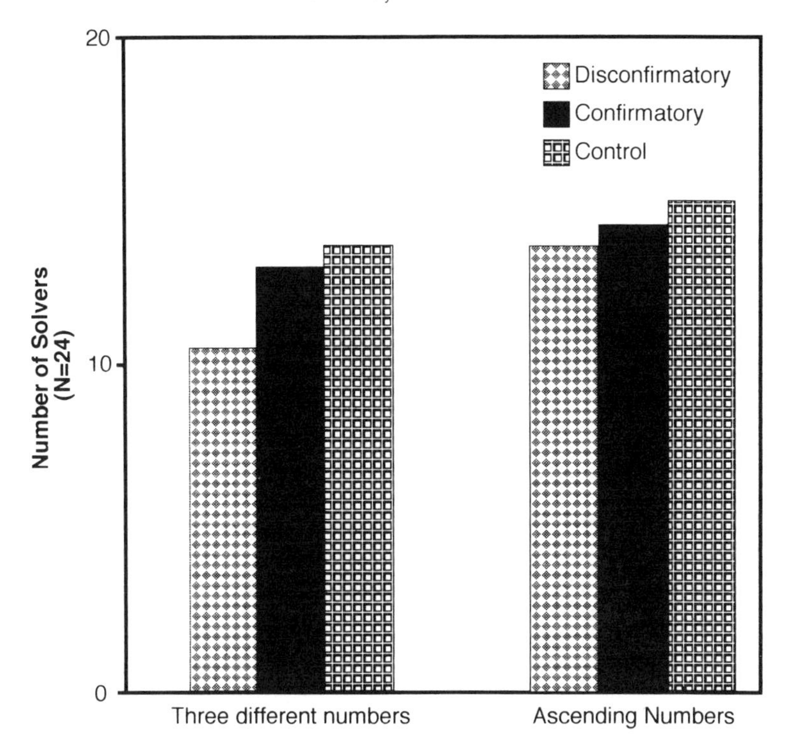

FIGURE 3-3 NUMBER OF SOLVERS WHEN RULES REVERSED

that rule, disconfirmatory subjects tried to obtain more incorrect triples on it, but they failed to do so and they therefore solved the rule significantly less often than did subjects in other conditions. Results for the "ascending numbers" rule were somewhat different, however. Disconfirmatory subjects predicted and obtained similar proportions of incorrect triples whether this rule was encountered first or second, but five fewer disconfirmatory subjects solved this rule when it was encountered second. Control subjects predicted and obtained a higher proportion of incorrect triples in this second study, and consequently solved the "ascending numbers" rule twice as often as did control subjects in the first study. Results for confirmatory subjects were almost exactly the same across both studies.

My first Eleusis and 2-4-6 studies suggested that a disconfirmatory heuristic is valuable when "God" can't tell you whether your rule is right. Did these latest results falsify the value of a disconfirmatory heuristic? No—they restricted its generality. Clearly, the nature of the rules and the order in which subjects attempt to solve them are important variables that affect the circumstances under which disconfirmation is effective.

I knew these caveats wouldn't matter to Popper. He was concerned only with the logic of justification; if falsification was the only logical way to test hypotheses, then that is what scientists ought to do, whether or not they or subjects or anyone else ordinarily were able to do it. But I felt intuitively that this sort of research did matter. If falsification failed under ideal laboratory circumstances, then how could it possibly work in more complicated situations like the "Chariots of the Gods" problem that motivated my research? Popper's prescriptions might represent an impractical ideal.

3.3. PERILS OF PUBLICATION

At this point I could only speculate as to why I got these results, but they seemed important enough to publish. I was still a Popperian in attitude; these results appeared to disconfirm, or at least raise problems with, my own earlier results. Perhaps if I published them in a journal some more-theoretically-inclined psychologist could explain them. I was aware, of course, of the prejudice against negative results, but I sent the manuscript to the *Quarterly Journal of Experimental Psychology* anyway. They had published my first paper on the 2-4-6 task; their audience would presumably be interested in a study which limited the generalizability of results from an earlier study.

My manuscript was reviewed thoroughly and promptly by two referees. One concluded that, "Whilst this study adds interesting qualifications to the authors' earlier findings, and would have made a very appropriate third experiment in the original paper, it hardly merits publication on its own. Apart from the relatively minor empirical contribution that this paper makes to the literature, I was also disturbed by the complete lack of any theoretical motivation for the study, and the absence of any attempt to discuss the results in a theoretical framework: the Discussion is almost entirely descriptive and anecdotal." The referee had a point. My graduate program in psychology had emphasized methodological rigor, somewhat at the expense of theory; hence, I tended to focus on detailed analysis of subjects' responses, rather than on making connections to theoretical issues. I had assumed it was enough that this study raised problems for my own previous research on disconfirmatory instructions.

The second referee objected to the comparison between experiments. This is a standard methodological concern: if one compares results from experiments conducted at different times, then the time difference might account for any difference in results. I thought I had answered that objection by showing that the subjects were similar and that a careful look at their reasoning processes supported my conclusion that there was negative transfer. This referee also wanted more references to scientific thinking, a criticism that pleased me enormously. He was somewhat more positive overall: "I would suggest that the authors either include the experiment with other research in a more substantial paper, or else rewrite a much shorter, clearer and less esoteric version of the present paper."

The *Quarterly Journal of Experimental Psychology* rejected this manuscript; I found I agreed with their decision. The reviewers' criticisms were constructive: not only had they uncovered important problems, they had pointed me toward the solution—add more theory and another experiment. They also confirmed my growing awareness that science is not a naively Popperian enterprise. Negative results are only interesting when they point to some positive alternatives.

3.4. THE SEARCH FOR TWO RULES

In order to understand the task-reversal effect, I felt I needed to have a better feel for subjects' performance on the "three different numbers" rule. Note that I was proceeding intuitively here, not deducing falsifiable hypotheses from a theory as a good Popperian should. If disconfirmatory instructions failed to improve subjects' performance, perhaps I could go back to Tweney et al.'s (1980) classic study (see 2.4) and adapt the procedure they used in their fourth experiment.

3.4.1. Substituting DAX and MED for Right and Wrong

Wason suggested to Tweney and his collaborators that they alter the 2-4-6 task so as to transform disconfirmation from a search for negative information to a search for information belonging to a different classification or rule. Instead of labeling triples right or wrong, Wason suggested they be labeled DAX or MED—a neutral, meaningless designation. Now subjects trying to falsify would not have to try to get triples "wrong"; all they needed to do was find triples that were instances of a different category.

When Tweney et al.'s subjects were run under DAX-MED conditions, their performance improved dramatically: 60 percent of the subjects solved the DAX rule on the first announced guess, and all but 15 percent solved it eventually. DAX-MED conditions made a significant improvement in performance in a situation where disconfirmatory instructions had not.

3.4.2. DAX-MED and Different Numbers

It occurred to me that it might make sense to see if this DAX-MED design would produce a similar improvement in performance on the "three different numbers" rule. I was here following a kind of "when in doubt, replicate" strategy. Given the differences in my rule and design, such a study would represent an important generalization of Tweney's results. It would also help me understand his methods; there is nothing like attempted replication to make one intimately familiar with another researcher's design.

Therefore, my wife, a student, and I (Gorman, Stafford and Gorman

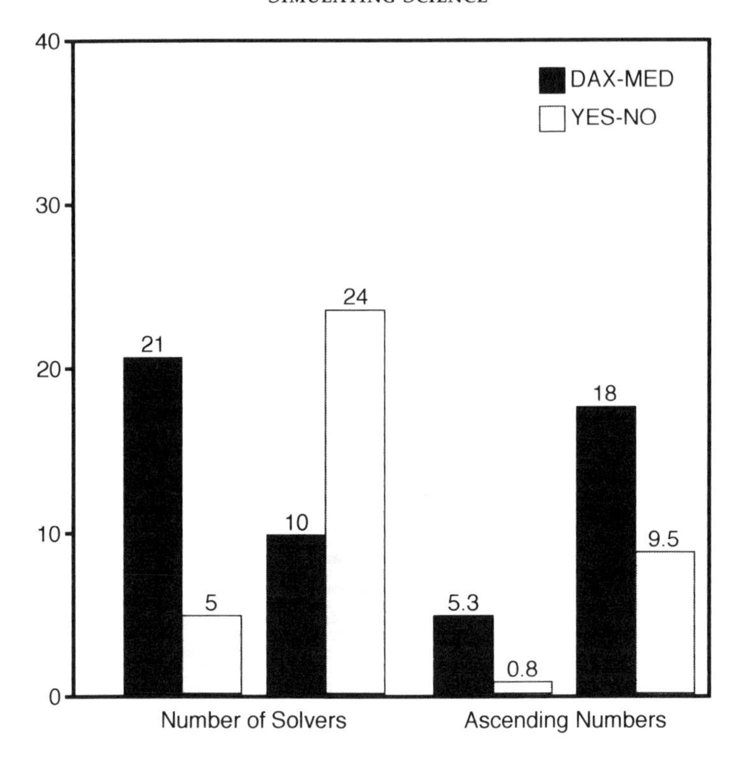

FIGURE 3-4 DIFFERENCES BETWEEN DAX-MED AND YES-NO CONDITIONS ON THE "THREE DIFFERENT NUMBERS" RULE

1987) compared subjects' performance on the "three different numbers" rule under two conditions:

(1) DAX-MED: Instead of being classified as correct or incorrect, each triple was classified as DAX or MED. "Three different numbers," was the DAX rule, which meant that the MED rule was "two or more numbers the same." The triple 2,4,6 was given as an instance of the DAX rule.

(2) Control: These subjects were run under the usual "Y" or "N" condition, with Ys corresponding to correct triples and Ns corresponding to incorrect.

Differences between conditions were dramatic. DAX-MED subjects solved the rule significantly more often than did control subjects, predicted and obtained a significantly higher proportion of incorrect triples, and proposed significantly more triples (see Figure 3-4).

Tweney et al.'s (1980) effect for DAX-MED instructions was clearly replicated on the "three different numbers" rule. But why did DAX-MED work, especially in a situation where disconfirmatory instructions did not? Tweney et al. offered three explanations:

(1) Telling subjects that some triples are incorrect interferes with their

ability to integrate this information. But studies discussed in sections 2.3 and 2.4 (reported as Gorman et al. 1984 and Gorman and Gorman 1984) have shown that it is possible for subjects to make good use of negative information. In the Eleusis study, especially, disconfirmatory subjects were told to study incorrect cards to get clues about the rule.

(2) DAX and MED hypotheses can be processed separately. A good analogy here may be to some of Ryan Tweney's (1985) work on Michael Faraday's discovery of electromagnetic induction. Faraday experimented on two, complementary hypotheses separately: one said that electricity could be induced from an electromagnet and the other said that electricity could be induced from an ordinary magnet. Confirmations of the first hypothesis came almost immediately, but the second hypothesis was far harder to confirm. As Tweney noted:

> Faraday seems to have played off an easily verified hypothesis against one that was not so encouraging at the start. So, when things were discouraging on the ordinary magnet side, he switched to electromagnets, to pursue further knowledge about the effect he knew how to produce. As he gained expertise, he was able to refine his attempts with ordinary magnets until he succeeded. . . .[This] reconstruction is plausible from a cognitive standpoint because similar 'dual hypothesis' strategies have been shown to be effective in laboratory studies of problem solving. Working with two closely related hypotheses helps because disconfirming evidence for one leaves the investigator with the other hypothesis as a focus of effort. Platt's advocacy of 'strong inference' is based on the similar point that to pursue several alternate hypotheses is an effective strategy. (Tweney 1985:204)

The DAX-MED experiment is the "dual-hypothesis" research referred to in the above quotation. Tweney sees a similarity between DAX and MED hypotheses and Faraday's "electricity from electromagnets" and "electricity from ordinary magnets" hypotheses. But the DAX and MED rules are mutually exclusive alternatives whereas Faraday's two hypotheses can almost be considered variants of the same hypothesis. Therefore, in Faraday's case success with one hypothesis suggests that the other has promise as well, whereas success in confirming the DAX rule does not guarantee success in discovering MED.

Tweney's larger point is that the DAX-MED design may encourage subjects to follow Platt's strong-inference heuristic by forcing them to keep at least two hypotheses in mind at all times and, like Faraday, to switch back and forth between them. But Platt is really talking about competing explanations for the same result, whereas the DAX and MED hypotheses are separate explanations for positive and negative results.

To see if subjects in the DAX-MED condition were really following a "hypothesis-switching" strategy, we asked subjects to predict whether each triple they proposed would be a DAX or a MED. A comparison of DAX-

MED and control conditions provided some support for Tweney's position: DAX-MED subjects switched from predicting one rule to predicting the other 35 percent of the time, whereas control subjects switched from predicting Y to predicting N only 26 percent of the time. This difference is statistically significant, but it is partially explained by the fact that there were significant differences between conditions in terms of proportion of incorrect or MED triples predicted and obtained. In fact, five control subjects proposed only triples they thought would be Y. In other words, control subjects simply proposed fewer triples they thought would be N, and therefore switched from one kind of triple to the other less often than did DAX-MED subjects.

(3) Tweney et al.'s final explanation is that the DAX-MED manipulation alters the logical structure of the task. Tweney does not spell out the implications of this alteration, but new research on the selection task helped us focus on how the DAX-MED design changes the way subjects mentally represent the 2-4-6 task. This analysis was also prompted by a referee's report.

3.4.3. Perils of Publication, Revisited

I sent the DAX-MED results to the *QJEP* in combination with the reversed-rule experiment they had rejected earlier. I hoped that combining a negative result with a positive one would help me get both the disconfirmatory and the confirmatory news out. Two referees were more favorably disposed to this new paper. One was concerned with the old two-experiment problem and wanted me to include both orders of rules in the same experiment. In a letter to the editor I argued that the samples and conditions were the same in both studies: "The only difference is that one study was run a short time after another. I would be willing to concede in the paper that I have not eliminated the possibility that the elapsed time between studies could somehow have accounted for the difference in results. But I do not think that the elapsed-time variable justifies rerunning both studies, as that would be enormously time-consuming and yield little new information" (letter from M. E. Gorman to N. Harvey, October 2, 1985).

As support for my argument, I pointed out that the second referee had not objected to the cross-study comparison. His concerns related to the nature of confirmation and disconfirmation. I had argued that the DAX-MED manipulation served to make disconfirmatory information more salient. This reviewer objected: how could I be sure subjects who tried to obtain MED triples weren't simply pursuing a confirmatory strategy with respect to the MED rule? He suggested that the DAX-MED manipulation was effective because it restructured the problem. This tallies nicely with Tweney's third possible explanation for his DAX-MED results. I felt immediately that this reviewer was onto something. It was left to me to figure out what.

3.4.4. Mental Representations and the Selection Task

Johnson-Laird and Wason (1970) developed a "reduced-array selection task" (RAST), in which subjects were asked to determine the truth of a sentence like "All the triangles are black" by inspecting either black or white cards, one at a time. Subjects knew there were fifteen cards in each color, and that each card would be either a triangle or a circle.

This task resembles the classic Popperian problem of determining whether "all swans are white," except that the universe of choices is restricted to two shapes and a small number of black or white cards. The logical solution is to inspect only the white cards; if one white card is a triangle, then the rule is falsified, even if every black shape is a triangle.

Subjects preferred disconfirmatory evidence, selecting over five times as many white shapes as black shapes. Why did a kind of "confirmation bias" appear to operate on the traditional selection task but not on this reduced-array version?

To find out, Wason and Green (1984) induced a unified or a disjoint mental representation on the RAST by the way they presented the materials. They altered the task slightly, presenting subjects with eight black and eight white cards whose lower halves were concealed; subjects knew that these concealed parts contained either circles or triangles. Subjects in the unified condition were asked to prove true the description "whenever there are triangles they are on black cards," and subjects in the disjoint condition were asked to prove true that "whenever there are triangles below the line there is black above it." Subjects in the unified condition realized more quickly that the only cards capable of proving or disproving the rule were white ones. Wason and Green concluded that "the subjects using a unified mental representation, compared to those using a disjoint representation, are less inclined to be fixated on the color used in the test sentence. The uncited color will become more salient, and hence be sampled more frequently, either overtly or covertly" (p. 609).

The positive effect for familiar forms of the selection task could be explained in a similar way: a familiar form, e.g., the drinking age in Florida, invokes a unified mental representation, one which makes the role of the disconfirmatory alternative "obvious." As Wason and Green concluded, "A unified mental representation of material does not seem to demand the retrieval of the relevant truth table nor any conscious inferential steps. The processes of solution are not open to introspection, because understanding of the problem seems to be immediate" (p. 608).

Similarly, one could argue in the Tweney and Yachanin (1985) experiment (reported in 1.2.1) that the words "test" and "determine if true or false" invoked different mental representations on the part of scientists; again, the true-or-false language made it "obvious" that the disconfirmatory alternative should be checked. Future research might be directed at determining whether this true-or-false language created some kind of "unified mental representation" or simply allowed scientists to access the appropriate logic.[6]

Studies of the differences between experts and novices in a variety of areas illustrate the importance of experience in forming appropriate representations. For example, the classic studies of the difference between chess masters and novices show that when a chess master is shown a board position briefly he or she remembers it much better than does a novice unless the position is scrambled and could not have resulted from an actual game. The chess master's superior memory is linked to action: "recognition of a pattern often evokes from memory stored information about actions and strategies that may be appropriate in contexts in which the pattern is present. A chess master recognizing that one of the files on the board is open—free of pieces—realizes immediately that one of his rooks might be moved to the foot of the file" (Larkin, McDermott, Simon and Simon 1980:1336). Perhaps novice performance on chess problems could be improved by a kind of RAST procedure in which the board was reduced to a few pieces in a critical relationship that could form a "unified mental representation" for a novice.

Just as the chess master knows immediately where the strengths and dangers of a particular board position lie, so the expert scientist forms a representation that tells her where to look for confirmatory or disconfirmatory information. Larkin et al. (1980) showed that expert/novice differences on complex, textbook physics problems resemble those of chess masters and novices, and discussed the kinds of abstract visual representations formed by experts—representations that, in turn, cued the appropriate set of equations (see 8.1.3).

Solving a textbook physics problem is different from testing a sophisticated theoretical hypothesis. But consider Einstein, who may have formed a kind of "unified mental representation" when he developed his theories of Special and General Relativity. His famous 1905 paper starts "with a curious question: why is there in Maxwell's theory one equation for finding the electromotive force generated in a moving conductor when it goes past a stationary magnet, and another equation when the conductor is stationary and the magnet is moving?" (Holton 1973:197–198).

Similarly, his theory of General Relativity focused on resolving asymmetries. One such paradox is contained in the case of an observer inside an elevator. How could he use experiments with falling objects to determine whether it was stationary or being pulled upward with a constant acceleration? The effects of gravity and constant acceleration would be the same—even, as Einstein pointed out, if one shone a beam of light through a hole on one side of the elevator: it would be slightly curved as a consequence either of gravity or of constant acceleration (Einstein and Infeld 1938). So why use different explanations for the two situations?

In both Special and General Relativity, Einstein sought to unify the laws of physics in a way that removed asymmetries and made them invariant for all observers. But did this goal of unification translate into any unified representations? Again, it is hard to say, because Wason and Green's use of the term is linked to a specific methodological permutation of the selection task, but the moving magnets and elevators above suggest that Einstein did, at least, formulate novel representations that unified phenomena. Note the way in which these representations, for Einstein, were often situations he could visualize; we will have more to say about this sort of representation when we discuss mental models in chapter 10 (see note 24).

3.4.5. Falsification and Representation

One implication of Wason and Green's study is that changing subjects' representations may have a more powerful effect on their performance than disconfirmatory instructions. This might explain why the DAX-MED design improved subjects' abilities to solve the 2-4-6 task when disconfirmatory instructions did not. The DAX-MED design emphasizes the role of non-DAX information; subjects immediately realize that the key to the problem is discovering the MED rule. As one subject said on a follow-up questionnaire about strategy, "I tried to change from the first triple in all ways to find out what the MED rule was." Furthermore, when they discover their first MED instances, they do not stop until they have made some attempt to identify the limits of the MED rule. The subject quoted above found his first MED instance on triple ten—8,8,8—and after proposing 1,1,1, concluded that the MED rule was "all the elements are the same number." Before he wrote down the DAX rule he paused and tried 1,1,2, which he predicted would be DAX but which turned out to be MED. Then he wrote down the correct rules for both DAX and MED.

Was this subject following a confirmatory or a disconfirmatory strategy? He was trying to find positive instances of the MED rule; these instances had the additional effect of determining the limits of the DAX rule. In effect, a strategy that was confirmatory with respect to DAX was disconfirmatory with respect to MED and vice versa. Consider his final triple, 1,1,2. He predicted it would be DAX; when it was MED, it disconfirmed his MED rule. Three DAX-MED subjects failed to propose this kind of additional triple and stopped with a DAX rule of "three different numbers" and a MED rule of "three numbers the same."

Thus, Tweney et al.'s third alternative comes the closest: the DAX-MED manipulation changes the logical structure of the task, from one in which subjects seek to disconfirm a single rule to one in which they try to confirm two mutually exclusive rules. This change in structure is reflected in the way subjects represent the task. Consider what happened when a DAX-MED subject encountered the triple 0,0,0, as two did. Both of these subjects used the fact that this triple was MED as a clue and tried a variety of similar triples to explore the MED rule. In contrast, the one control subject who encountered 0,0,0 stopped and concluded that it was an exception; his final guess was "three consecutive integers, each multiplied by the same number (but not 0)." He was looking for a single rule that might have exceptions; DAX-MED subjects represented the task as a search for two distinct rules. Further evidence for this view comes from the fact that no DAX-MED subjects thought the DAX rule could be "any three numbers"; indeed, the one subject in this condition who failed to find a single MED instance realized he had simply failed to find the rule, and was still convinced that there was one. In contrast, six control subjects decided the rule was "any number" because their exhaustive search failed to turn up a single N.

To summarize, DAX-MED subjects represented the task as a search for two mutually exclusive, complementary rules, whereas Y-N subjects represented the task as a search for a single rule which might or might not have exceptions. As in Wason and Green's RAST problem, subjects working on the DAX-MED version of the 2-4-6 task immediately recognize the importance of the non-DAX instances and know that they are searching for a rule that governs them. In the Y-N version of the same task, subjects are unable to obtain sufficient information about incorrect instances; when they do get triples wrong, they fail to explore how this negative information could fit into a pattern, and instead regard these instances as exceptions. Naturally, there were individual differences, but it is impressive how effective this manipulation was in altering task representation, which in turn affected strategy. DAX-MED instructions probably do not create a more "unified" representation, as subjects represent the task as a search for two rules. However, this study, in combination with Wason and Green's work, suggested the importance of problem representation, even on very abstract tasks.

The *Quarterly Journal of Experimental Psychology* accepted an argument along these lines and published the reversed rules and DAX-MED experiments together in the same article. I was learning that sequencing experiments was an effective strategy for getting published; it meant fewer publications, but one could present a coherent research program, rather than an isolated study.

3.5. THE RELATIONSHIP BETWEEN REPRESENTATION AND RULE

Shortly after this article was published, I read an article by Klayman and Ha (1987) which helped put the relationship between mental representations and heuristics such as confirmation and disconfirmation into perspective. They discussed the interaction between the nature of the problem and when positive (confirmatory) and negative (disconfirmatory) test heuristics should be employed. They preferred the terms *positive* and *negative* because whether a heuristic could be called confirmatory or discon-

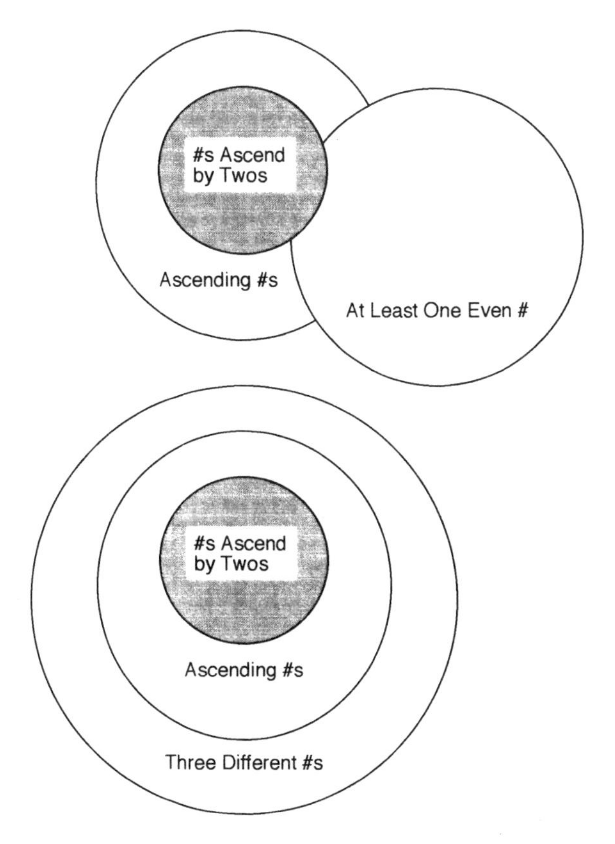

FIGURE 3-5 RELATIONSHIP BETWEEN HYPOTHESIZED AND TARGET RULES

firmatory depended on the relationship of the target rule to the hypothesized rule. "Target rule" corresponds to the experimenter's rule and "hypothesized rule" corresponds to the subject's current hypothesis. On Wason's original version of the 2-4-6 task, the triple 2,4,6 makes it likely that subjects will begin seeking a hypothesized rule which is embedded in the target rule, i.e., the target rule is more general than hypothesized rules like "numbers go up by twos" and "ascending multiples." In this situation, a negative test heuristic leads to disconfirmation because a number of the instances which lie outside of the hypothesized rule turn out to be correct (see Figure 3-5).

In the case of the "at least one even number" rule, hypothesized rules like "numbers go up by twos" or "ascending numbers" would overlap the target rule. A subject holding either of these hypotheses and pursuing a confirmatory heuristic might propose the triple 1,3,5, which would be incorrect, disconfirming the hypothesis. In other words, a confirmatory or positive test heuristic in this situation is just as likely to produce falsifications as a disconfirmatory heuristic would be.

"Three different numbers" is a target rule even more general than Wason's original one; if a subject takes "ascending numbers" as his or her hypothesized rule, then he or she must follow a negative test heuristic to discover that "three different numbers" is the target rule. Therefore, a disconfirmatory or negative heuristic makes logical sense on this rule. The problem is, subjects don't know where to look for disconfirmatory information; they can't seem to visualize alternatives to their hypothesized rules.

What happens when a subject has already completed a task involving the more general target rule and now must find a second rule that is embedded in the first? Consider moving from "three different numbers" to "ascending numbers." Hypotheses generated while working on the first rule may be carried over to the second—and may be more general than the second target rule. In a case where the target rule is narrower than the hypothesis suggests, it makes sense to follow a confirmatory or positive test heuristic. For example, one disconfirmatory subject in Experiment Two who solved the first rule began the second with the same hypothesis. She tried several disconfirmatory triples, all of which had two or more numbers the same, and when all were incorrect, concluded that the rule was "the three numbers must be different." What she failed to do was try some confirmatory triples, e.g., one with descending numbers, which would have showed her that her hypothesized rule was more general than the target rule.

In other words, in a situation where the first rule is more general than the second, successful subjects have to pursue a confirmatory heuristic at least part of the time on the second. Disconfirmatory subjects, therefore, were instructed to follow an ineffective heuristic when the "ascending numbers" rule followed the "three different numbers" rule.

Thus, the relationship between target and hypothesized rules may explain the failure of disconfirmatory instructions to improve performance on the "ascending numbers" rule when it was in the second position. Can Klayman and Ha's analysis also explain performance on the DAX-MED version of the task? Unfortunately, their analysis falls prey to the same objection as a confirmatory-disconfirmatory analysis: when we say that subjects seek MED instances, that is a positive heuristic with respect to the MED rule and a negative one with respect to the DAX. The key here is not just the relationship between hypothesized and target rules; instead, the DAX-MED manipulation transforms the way in which subjects frame the whole task. Instead of one rule that might have exceptions, they now know that they are seeking two mutually exclusive rules.

So, a more complex analysis would include not only the subject's hypothesis and target rule, but also how the subject represents the target rule. If a subject represents the target as a rule with a few idiosyncratic exceptions, then whether a confirmatory or a disconfirmatory heuristic is effective will depend on whether those exceptions are more specific or more general than the subject's hypothesis. In effect, the best strategy in this situation is a combination of confirmation and disconfirmation, though a completely exhaustive search is impossible.

If, however, one represents the target rule as consisting of two, mutually exclusive categories, the best heuristic is to seek positive instances of both categories, then inspect the borders. This "border inspection" will involve both confirmation and disconfirmation of one's hypotheses about the rules and one's representation of the relationship between the rules. Consider, for example, the possibility that the rules might overlap, or that they might both exclude certain instances that lie outside both. The DAX rule might be "ascending numbers" and the MED rule might be "odd-even numbers must alternate," in which case certain triples like 1,4,7 would be instances of both DAX and MED and certain others like 1,3,1 would be instances of neither. Clearly, the relationship between confirmation and disconfirmation becomes far more complex when one reframes the 2-4-6 and related tasks as a search for two or more categories rather than a search for a single rule.

3.6. REPRESENTATIONS AND HEURISTICS

The research cited so far has demonstrated that whether a disconfirmatory heuristic is effective depends on the relationship between hypothesis and target rule, and also on how the subject or scientist represents the nature of the task. On a very general rule like "three different numbers," it is not sufficient for subjects to know that they must disconfirm; they must know where to look for disconfirmatory information and be able to formulate appropriate alternate hypotheses. The DAX-MED manipulation helps subjects represent the task in a way that facilitates the search for alternatives.

These findings highlight the critical relationship between mental representations and heuristics. Even earlier results which demonstrated the importance of a disconfirmatory heuristic were dependent, in part, on how instructions affected subjects' representations. In Gorman et al. (1984) subjects were not told just to disconfirm; they were told to try to get cards wrong and to focus especially on incorrect cards. In other words, these instructions told subjects not only what heuristic to apply, but where to look for information relevant to that heuristic. On this Eleusis task, hypothesized rules like "ascending diamonds" tended to be embedded in target rules like "odd and even cards must alternate"; therefore, it was helpful to represent the task as a search for incorrect cards.

Support for this interpretation of Gorman et al.'s results comes from research I became aware of recently using groups on a card task very similar to Eleusis. Laughlin and Futoran (1985) compared the performance of groups and individuals under three instruction conditions: disconfirmatory, confirmatory, and control. They replicated my first Eleusis results (see 2.2.3) except that interacting groups in the control condition performed better, because "Instructions to select cards to confirm or disconfirm hypotheses are an additional requirement that may have added to the complexity of an already difficult task, and they may have interfered with the basic objective of inducing the rule by both confirming and disconfirming card selections as seemed appropriate at various stages of the problem" (p. 611). Why did they obtain no instruction effect when Gorman et al. did? First, Laughlin and Futoran included several rules, some of which might have been more general than a subject's initial hypothesis, and others more narrow; therefore, when they lumped all these rules together for analysis, positive effects for confirmatory and disconfirmatory instructions may have canceled each other out. Second, they provided few details concerning the examples they used to illustrate confirmatory and disconfirmatory instructions, and examples are critical in forming an appropriate procedural representation of the heuristic being applied.

3.6.1. The Declarative/Procedural Distinction

Anderson (1983) distinguished between declarative knowledge, or knowledge of "what," and procedural knowledge, or knowledge of "how to." Laughlin and Futoran's instructions to disconfirm may have given subjects a kind of declarative knowledge of what heuristic they were supposed to follow but not an appropriate procedural knowledge of how to follow it. Examples and practice are necessary to transform declarative knowledge into procedural.

Anderson claims that "The procedural learning mechanisms, which are gradual and inductive, contrast sharply with the abrupt and direct learning that is characteristic of the declarative domain" (1983:35). He uses the example of learning how to drive a car with a stick-shift: first you have to know where the gears are and what they do, then you have to gradually transform that declarative knowledge into procedural. Broadbent (1987) has suggested that one can learn procedures in an implicit manner without first learning them in a declarative form.

The major point for our purposes is that instructions need to be transformed into a kind of procedural representation before they can be implemented. Furthermore, effective use of procedures depends on the formation of an appropriate task representation (Singley and Anderson 1989). In my Eleusis and 2-4-6 studies, I found I had to use detailed examples illustrating strategies and emphasize studying correct cards in the case of confirmatory and incorrect cards in the case of disconfirmatory. This helped subjects transform their declarative heuristic into a procedure they could apply. Furthermore, the relationship between subjects' initial hypotheses (e.g., "ascending diamonds") and the target rule (e.g., "odd and evens alternate") made the disconfirmatory procedure far more effective. According to Laughlin and Futoran, they checked carefully to make sure subjects were following their strategy instructions, but they provide no detailed examples of subjects' processes, nor could they have been aware of more recent work on the relationship between representation and rule. Reading their study several years after it had been done made me realize how much more sophisticated our understanding of confirmation and disconfirmation had become.

3.6.2. A Hierarchy of Heuristics

The 2-4-6 task and Eleusis are deliberately designed so that prior knowledge will play as little role as possible, whereas the working scientist possesses a large store of expertise and information which he or she brings to bear on every problem (see Larkin 1983). Therefore, tasks like Eleusis highlight the role of what Langley, Simon, Bradshaw, and Zytgow (1987:11–12) call weak heuristics:

> We can picture a hierarchy of heuristics, with the top levels consisting of very general algorithms that require little information about the task domain to which they are applied and are correspondingly applicable to a great many domains. As we proceed downward in the hierarchy of heuristics, we make more and more demands on information about the task domain, and there are correspondingly fewer domains to which the heuristics apply—they become more and more task specific as we descend through the hierarchy. Since the more general heuristics operate on the basis of less information, we may expect them to be less selective, and hence less powerful, than heuristics that make use of more information about the task structure.

Weak heuristics are particularly useful when one approaches a novel problem that lies outside one's traditional expertise, but they may also be used in combination with strong heuristics. For example, Kulkarni and Simon (1988) argued that the chemist Krebs combined weak and strong heuristics in his discovery of the ornithine cycle (see below, 9.2.2).

3.6.3. Chariots of the Gods Revisited

To illustrate the relationship between representations and heuristics, let us go back to an example discussed in the Introduction. Part of my skepticism toward Chariots of the Gods came from a kind of disconfirmatory heuristic—I had been taught procedurally, by dozens of examples, to think critically. But part of it came from the way I represented what constituted appropriate research in archaeology and anthropology and from declarative knowledge relevant to these topics, based on my beginning courses in these areas. Again, it is not sufficient to know that one must think critically; one must have sufficient background knowledge and be able to represent the problem in a way that enables one to identify the weaknesses in a hypothesis like Chariots of the Gods.

Years later, when I encountered the Alvarez hypothesis about the extinction of the dinosaurs, my disconfirmatory heuristic was supplemented by declarative knowledge that indicated the dinosaurs had not disappeared suddenly, but over a much longer period of time than usually thought. I also represented the evolutionary problem as requiring a solution that emphasized its gradual and continuous nature. Had my declarative knowledge emphasized the extinction as sudden, and my rule representation included the possibility that geological catastrophes were major sources of evolutionary change, I would have been less critical of the Alvarez hypothesis when it first appeared.

3.6.4. Future Research on Representations and Heuristics

As noted earlier, subjects working on laboratory simulations of scientific reasoning do not have the organized prior knowledge or the strong heuristics of the expert; these tasks are ideal for studying the relative value of "weak" heuristics like confirmation and disconfirmation. But even on these tasks, heuristics do not operate in a vacuum. Subjects do gradually build declarative knowledge, based on the results of their experiments or trials, and this knowledge suggests where to look for new confirmations or disconfirmations.

Furthermore, subjects quickly form a representation of what sort of rule they are seeking. Is it a general rule, or a specific one? Will it be a single rule with exceptions, or a rule that clearly delineates two types of triples?

Future experimental research could focus on different possible relationships between heuristics, hypotheses, and rule representations. Klayman and Ha (1987) assert that a confirmatory or "positive test" heuristic is most valuable when the target rule is embedded in the hypothesized rule, and a disconfirmatory or "negative test" heuristic is most valuable when the hypothesized rule is embedded in the target rule. Obviously, these target/hypothesis relationships depend not only on the actual relationship between rules but on the subject's representation of them.

Variations on the DAX-MED design would be ideal for this purpose. For example, what if the DAX and MED rules were not mutually exclusive but could overlap and/or have triples that were excluded by both? DAX could be "ascending numbers" and MED "odd and even numbers must alternate." In this case, a triple like 1,4,9 would be both DAX and MED and 3,3,3 would be neither. Or the DAX rule could be embedded in the MED rule and both exclude some triples: DAX might be "ascending numbers" and MED "three different numbers." In this case, every DAX would also be a MED, and triples like 3,3,3 would be excluded. The possible permutations are endless. The point would be to study subjects' representations of these rules and the relative value of confirmatory and disconfirmatory heuristics. When DAX and MED rules overlap and exclude, it is possible to talk about heuristics that test the limits of both (disconfirmatory) and explore common ground (confirmatory), as well as heuristics that are disconfirmatory and confirmatory with respect to each rule. Will subjects, in the absence of any hints, quickly realize that the rules overlap and exclude? Will they focus on finding the limits of each rule, or of both?

Another area for exploration is the relationship between heuristics and representations when there is error in the data. On Eleusis and the 2-4-6 task, subjects know every result is reliable. What would happen if they knew some of these results could be erroneous? Would they use errors to immunize hypotheses? Klayman and Ha (1987) argue that a confirmatory or positive test heuristic is particularly useful when error is possible. Chapters 4 through 6 will describe a program of research designed to deal with these questions.

Clearly, even on abstract laboratory tasks, heuristics depend on representations. In chapter 10, we will come back to this issue and present a framework for studying the cognitive processes of inventors and scientists which emphasizes the relationship between heuristics and artifacts.

3.7. WHEN CONFIRMATION IS CRUCIAL

This chapter began with results which seemed to indicate that, even in the absence of feedback on hypotheses, training subjects to adopt a disconfirmatory heuristic does not always improve their performance. But this apparent "falsification of disconfirmation" can be explained: the usefulness of a disconfirmatory heuristic depends on (a) whether a subject has adequate procedural knowledge of how to disconfirm and (b) how the subject represents the relationship between the target rule and his or her hypothesized rule.

Popper regarded verification as a logically untenable strategy, but research indicates that it possesses considerable value. In situations where the subject's representation or hypothesis about the rule is more general than the actual rule, a strategy of seeking positive instances will lead to disconfirmation. Also, as Mynatt, Doherty, and Tweney (1978) showed, confirmation is an extremely useful strategy early in the inference process, when subjects and scientists are groping for a pattern.

Indeed, one might argue that the 2-4-6 task shows that confirmation is a useful strategy during the discovery phase of inference, but that at some point a switch to disconfirmation is essential to test a hypothesis. Tweney, Doherty, and Mynatt (1981) argued that one should confirm either early in the inference process and/or when the quality of the data is low; disconfirmation should be done under the opposite circumstances, when the data quality is high and/or when one already has a well-corroborated hypothesis. (We will have more to say about data quality in chapters 5 and 6, when we discuss error.) Michael Faraday seems to have followed this process (Tweney 1985).

But the context of discovery as traditionally used may be misleading here. Perhaps a better phase would be "the context of pursuit," as used by Laudan (1977) and others. In this intermediate phase, a scientist tries to determine which tentative "discoveries" are worth the time and effort of pursuing and justifying. Confirmation is a good heuristic for determining whether a tentative conjecture yields promising results.

So in conclusion, confirmation and disconfirmation can be blended into a "weak" or general heuristic as follows: try to confirm tentative hypotheses initially, to see if they are worth pursuing, then if a hypothesis appears well-corroborated, switch to disconfirmation. This implies that confirmation is not really a bias; indeed, it is a useful heuristic, especially when a scientist's hypothesis is more general than the law or relationship he or she is investigating. Naturally, a scientist does not know this "target rule" in advance, but if he or she represents the task as a search for a principle that is more specific than the current hypothesis, a confirmatory or positive test strategy makes sense. Consider the hypothesis proposed by advocates of SETI, the Search for Extraterrestrial Intelligence. Their hypothesis is that there is other intelligent life in the universe. They are searching for evidence of other civilizations by monitoring a variety of frequencies using radio telescopes. No amount of failure to detect signals indicating intelligent life is decisive, because it could be that the wrong frequencies are being monitored, or that the signal is in fact being received but not recognized, or that no civilization in our area exists or is sending. Conversely, a single successful detection of a signal from an intelligent civilization will confirm the hypothesis. In Klayman and Ha's terms, the hypothesis is more general than the target rule, in that the hypothesis does not specify under what precise circumstances a signal will be detected. Whether confirmation or disconfirmation is a sensible heuristic depends on how a scientist represents the relationship between his or her hypothesis and the law or principle he or she is seeking.

3.8. FALSIFICATION CRITICIZED: KUHN

At the same time as I was wrestling with the issue of mental representations in my experimental tasks, I was taking a second look at a philosopher I had read in college. Thomas S. Kuhn's (1962) *The Structure of Scientific Revolutions* created a revolution of its own. Philosophers often speak of "post-Kuhnian philosophy of science" to denote the sense of having crossed some kind of watershed with Kuhn. His work is widely cited outside of philosophy and history of science, and often misrepresented.[7]

Kuhn's major contribution is his distinction between two kinds of science. "Normal" science is what scientists do most of the time: they solve puzzles within the framework of what Kuhn calls a paradigm. Kuhn used this term in so many different ways (see Masterman 1970) in his *The Structure of Scientific Revolutions* that he later all but abandoned it (Kuhn 1970). His basic idea was that scientific specialists are trained in a particular way of looking at the objects of their inquiry through exemplars, which are "concrete problem solutions, the sorts of standard examples of solved problems which scientists encounter first in student laboratories, in the problems at the ends of chapters of science texts, and on examinations" (Kuhn 1970:272). Note the similarity between exemplars and mental representations. During periods of "normal" science, scientists seek problems that can be solved within the theoretical framework or paradigm suggested by these exemplars. If the result of an experiment is inconsistent with the paradigm, the result is questioned, not the paradigm.

But if enough of these inconsistent results accumulate, then the science enters a period of crisis and revolution, in which the reigning paradigm is overthrown and replaced by a new one. The choice of political metaphors here is deliberate; Kuhn saw paradigm shifts as similar to political or religious revolutions, except that scientists use empirical findings as weapons instead of swords.

Consider, for example, Newtonian mechanics. This paradigm dominated much of physics until the late nineteenth century, when what Kuhn calls anomalies began to appear. Anomalies are results that cannot be explained within the context of a paradigm; they bear a strong resemblance to Popperian falsifications, although anomalies do not eliminate a theory; instead, they precipitate periods of crisis. In the Newtonian case, the anomalies included the precession of the perihelion of Mercury: not only does Mercury rotate around the sun, but its elliptical orbit rotates as well, an effect not predicted by Newton's theory (Einstein and Infeld 1938). A similar anomaly that illustrated problems with reconciling Newton's mechanics with Maxwell's electromagnetic theory of light was the result of the Michelson-Morley experiment, which showed that the speed of light is the same in the direction of the motion of the Earth as it is in the opposite direction, indicating that light waves cannot be propagated in an "ether" through which the Earth is also moving. The Mercury anomaly was predicted by General Relativity; the Michelson-Morley result by Special Relativity. Interestingly, Einstein was aware of neither of these results when he formulated his theories (Einstein and Infeld 1938; Holton 1973). Michelson regarded his negative result as a failure, contra Popper's dictum (Holton 1973:66).

So although these anomalies may have helped create an atmosphere of crisis in the field—the Michelson-Morley result, in this case, far more than the precession of the perihelion of Mercury—they did not play an important role in the thinking of the man who developed the new paradigm.

Just as in politics revolution leads to a new form of government, so in science revolution leads to a new paradigm—Einstein replaces Newton, Copernicus replaces Ptolemy, etc. Not all scientists "convert" to the new paradigm; in fact, Max Planck somewhat pessimistically argued that "A new scientific truth does not triumph by convincing its opponents, but rather because its opponents eventually die, and a new generation grows up that is familiar with it" (quoted in Messeri 1988:92; contra Planck, Messeri found that younger geologists were slower to adopt plate tectonics). Kuhn uses the term "incommensurable" to describe the gulf between two paradigms; it requires a "gestalt-switch" in perspective to go from one to the other.

Here, Kuhn relies on some early Gestalt psychology experiments for an analogy. One in particular, by Bruner and Postman, involved displaying cards to subjects at brief exposures. Some of the cards were ordinary, and were identified immediately; others were anomalous, to use Kuhn's term, e.g., a black four of hearts. Initially, the subjects classified these anomalous cards into existing categories; e.g., the black four of hearts might be classified as the four of spades. But as the exposure time increased, subjects experienced some doubt and anxiety until finally most were able to classify the anomalous instances correctly—although some subjects were never able to recognize the anomalous cards. As one subject said, "I can't make the suit out, whatever it is. It didn't even look like a card that time. I don't know what color it is now or whether it's a spade or heart. I'm not even sure now what a spade looks like. My God!" (quoted in Kuhn 1962:63–64).

Like the subjects viewing playing cards in this experiment, scientists switching from one paradigm to another must learn to view their exemplars differently. Consider the case of Kepler (see Koestler 1963) trying to solve the problem of the orbit of Mars. Ptolemy and Copernicus both agreed that the orbits of planets had to be circular; this circular model, based originally on Aristotle's spheres, served as a kind of exemplar for Kepler as well. However, Kepler found it impossible to fit Tycho Brahe's data on Mars into a circular path and after several years of struggle realized that no combination of epicycles would work; all he had left was a shape he contemptuously referred to as "only a single cart-ful of dung": the ellipse. His disgust is evidence of the power of the circular exemplar, but eventually his elliptical orbits helped him discover a new set of harmonious relationships, embodied in his three laws.

Kuhn feels that he and Popper are close to agreement on what happens during these periods of extraordinary or revolutionary science. Popper does not talk about gestalt-switches and paradigms, of course, but he does argue that scientists ought to make risky conjectures and then seek to disconfirm them. This is a good strategy during periods of crisis. According to Kuhn, what Popper has missed is the significance of the long periods of normal science that lie between crises.

3.8.1. Heuristics in Normal and Revolutionary Science

Simon, Langley, and Bradshaw (1981) argue that normal scientists typically employ strong heuristics, whereas in periods of crisis, scientists are forced to rely more on weak or general methods:

> . . . scientific activity, particularly at the revolutionary end of the continuum, is concerned with the discovery of new truths, not with the application of truths that are already well known. While it may incorporate some expert techniques in the manipulation of instruments and laboratory procedures, it is basically a journey into unmapped terrain. Consequently, it is mainly characterized, as is novice problem-solving, by trial-and-error search. The search may be highly selective—the selectivity depending on much that is already known about the domain—but it reaches the goal only after many halts, turnings, and backtrackings. (p. 5)

Confirmation and disconfirmation are both weak or general heuristics that could, if one believes Simon et al., be most effectively employed during periods of crisis, when one is groping for new hypotheses and testing them in a trial-and-error fashion.

But of course the picture is really more complicated. One could argue that confirmation is the optimal weak heuristic during periods of normal science; scientists seek to solve problems or puzzles in ways that are consistent with the paradigm, and if the solution contradicts the paradigm, it is the result that is questioned, not the paradigm. Confirmation is coupled, in this case, with a wide range of stronger heuristics. Examples might be the instrumental techniques and methodological prescriptions Kuhn claims are part of the paradigm.

Disconfirmation becomes an optimal heuristic during the revolutionary phase. If Simon et al. are right, this sort of disconfirmation should take on more of a trial-and-error character; scientists should be more willing to conjecture boldly and seek refutations. Innovative strong heuristics might also be more likely to evolve.

3.8.2. Does Science Progress?

Kuhn and Popper disagree over the issue of progress. Both feel that science progresses, and that is something that distinguishes science from nonscience. But while Popper feels that science progresses toward the truth, that Einstein's theory of General Relativity is a truer picture of the universe than is Newtonian mechanics, Kuhn feels that Einstein merely explains more puzzles or problems than Newton. According to Kuhn (1970:265), in "some fundamental ways" Einstein's universe is closer to Aristotle's than to Newton's.

Kuhn also obviates the traditional distinction between the contexts of discovery and justification. According to Kuhn, theory rejection and acceptance are decisions made by a community of practitioners; to understand these decisions, one needs psychology and sociology, not just the logic of justification. Note that Kuhn is not a complete relativist. Logic and empirical results usually play a major role in theory choice. But logic is employed within a community to persuade or convert. Theory change cannot be understood as an abstract syllogism.

To summarize, Kuhn sees science as evolving from a preparadigmatic state to a normal science phase in which a paradigm emerges and becomes dominant. Eventually, anomalies force a state of crisis, a revolution occurs, and a new paradigm emerges. The social sciences are often cited as examples of preparadigmatic or protoscientific fields because there is an enormous debate over fundamentals in these areas. When the debate over fundamentals gives way to successful puzzle-solving, then these fields will be sciences, according to Kuhn. He believes that making specific, empirical predictions is one criterion for being a science; but if the predictions fail in a mature science, the result or experimenter is questioned, not the paradigm—until, at some undefined point, the anomalies become sufficient to precipitate a crisis.

3.9. FALSIFICATION REVISED: LAKATOS

Imre Lakatos tried to reconcile Popper and Kuhn. He began by distinguishing between three levels or types of falsification:

(1) Dogmatic falsification: This position corresponds to a naive version of Popper's conjectures and refutations. To be scientific, a theory must specify experimental consequences such that the theory will be abandoned if the results do not turn out as predicted. Science advances by making bold conjectures and (eventually) by refuting them with hard facts.

Lakatos contends that this is an overly simplistic view, for a variety of reasons, the most interesting of which is his contention that "exactly the most admired scientific theories simply fail to forbid any observable state of affairs" (1978:16). To support this argument, he tells a story about a scientist who calculates that the orbit of a newly discovered planet departs from Newtonian mechanics. Is Newton's theory falsified? No—as actually occurred in the case of Uranus, the scientist makes the assumption that another planet is causing the gravitational perturbations. If no other planet is discovered, the astronomer can invoke a variety of corollary assumptions to explain his or her result.

In other words, a hard core is never falsified. What is falsified is the conjunction of a theory and what Lakatos calls a "protective belt" of assumptions governing the relationship between specific events and the theory. One can always change one or more assumptions to protect the theory.

(2) Methodological falsification: The methodological falsificationist re-

quires that, to be scientific, a theory must specify under what grounds it can be rejected. The difference between methodological and dogmatic falsificationists is that the former recognize that rejection does not necessarily imply disproof—a valid theory could be rejected. The methodological falsificationist sets up in advance conventions or grounds for rejecting a theory, then sticks to those guidelines, even though it is possible that they might be wrong. The goal is to provide severe strictures so that only the fittest theories will survive.

Einstein's General Relativity is the classic example: he made specific predictions about the curvature of light in the sun's gravitational field; according to Popper, if results had varied significantly from Einstein's predictions, then his theory would have been wrong. However, when asked what he would do if actual measurements of curvature conducted by the distinguished British physicist A. S. Eddington during an eclipse contradicted his theory, Einstein said, "Then I would feel sorry for the dear Lord [Eddington]. The theory is correct" (Einstein, quoted in Holton 1973:234–235). Furthermore, Eddington (1934/1959:211) himself asserted that one should not "put overmuch confidence in the observational results that are put forward until they have been confirmed by theory."

This example illustrates why Lakatos was uncomfortable with methodological falsification: it does not square with "good decisions" in the history of science.[8] Why was the problem of Mercury's perihelion ignored for so long, when it was a well-corroborated falsification from a methodological standpoint? Why did Galileo and others embrace the Copernican system when the empirical evidence for Ptolemy's was just as good until Kepler developed his three laws? What the methodological falsificationist misses is the critical role played by alternative theories.

(3) Sophisticated methodological falsification: This approach takes into account the role of rival theories in falsification. A scientific theory is one that accounts for all of the evidence that supports its rivals and also makes some novel predictions that are corroborated. A consequence of this view is that "there is no falsification before the emergence of a better theory" (Lakatos 1978:35).

Auxiliary hypotheses can be invoked as adjustments to a theory if they lead to novel predictions; auxiliary hypotheses that merely "patch up" difficulties lead to the degeneration of a theory. In fact, auxiliary hypotheses that make novel predictions often lead to newer, more progressive theories. "It is a succession of theories and not one given theory which is appraised as scientific or pseudo-scientific" (Lakatos 1978:47).

Lakatos identifies this sophisticated methodological position with Karl Popper. It accounts, for example, for the success of General Relativity, which explained all the phenomena described by Newtonian mechanics and also made a set of novel predictions which were not refuted.

Lakatos went a step beyond sophisticated methodological falsification in developing his own set of heuristics for successful research programs.

A research program has a "hard core" that must be protected from falsification by a "protective belt" of auxiliary hypotheses. For example, according to Lakatos, in Newton's program the three laws of dynamics served as the irrefutable hard core; deducing the existence of Neptune from the perturbations in the orbit of Uranus is an example of how the belt of changing auxiliary hypotheses functions to protect the hard core.

Here we have the reconciliation of Kuhn and Popper. The hard core resembles Kuhn's paradigm; the protective belt is altered by activities that resemble puzzle-solving, none of which are permitted to affect the hard core. But there is Popperian evolution as well. In a progressive research program, modifications of the protective belt will constantly lead to successful novel predictions. In a degenerative research program, modifications of the protective belt will merely "patch up" difficulties with the hard core and will not succeed in predicting new results. There is no single "crucial experiment" that can determine which of two programs is closer to the truth, but, over time, one program will make more successful empirical predictions than the other. A research program is never falsified, but it can be replaced by a more successful rival.

For Lakatos, progress depends more on verification than on falsification:

> Our considerations show that the positive heuristic forges ahead with almost complete disregard of "refutations": it may seem that it is the "verifications" rather than the refutations which provide the contact points with reality. Although one must point out that any "verification" of the (n+1)th version of the programme is a refutation of the nth version, we cannot deny that some defeats of the subsequent versions are always foreseen; it is the "verifications" which keep the programme going, recalcitrant instances notwithstanding. (Lakatos 1978:52)

Of course, what counts as a confirmatory result for one program often counts as a disconfirmatory result for its rival.[9]

3.9.1. Mental Representations as Hard Cores

Results from experiments described in this chapter illustrate how representations can become "hard-core" assumptions that resist disconfirmation. Consider one of the subjects who solved the "three different numbers" rule before he encountered the "ascending numbers" rule (see section 3.2). He noted that triples like 9,10,8 and 10,20,19 were wrong on the second rule, whereas they had been right on the first, and so he proposed that the rule was "Any combination of numbers such that the middle number isn't larger than the two outer numbers." Note that his mental representation, derived from his experience on the first task, dictated the type of rule he was looking for. He went on, in good Lakatosian fashion, to test the novel predictions made by this rule, proposing 6,7,7, a triple that should be right according to his hypothesis. It was wrong.

Instead of abandoning his "research program," he merely modified his guess, adding the assumption that "the outer number can't be equal to the center number, either." He then continued to test novel predictions of his program, proposing the triple 8,7,8, which should have been right if his hypothesis were right. It was wrong. At this point, he was faced with the classic Kuhnian anomaly: the middle number was not larger than or even equal to the two outer numbers and the triple was correct. But instead of abandoning his research program, he simply ignored the disconfirmatory result and kept on trying other confirmatory triples, i.e., ones in which the inner number was larger than the two outer numbers. In Lakatosian terms, he embarked on a "degenerating research program."

This subject looks like a classic case of confirmation bias; he behaved irrationally by Popperian standards. From a Lakatosian or Kuhnian perspective, however, his performance makes sense: since he had no other hypothesis, he could not abandon the hard core of his research program. Why had he fixed on this idea? Because on the first task he had developed a heuristic of looking for ways in which two numbers could *not* appear in a triple. He was not prepared to abandon this successful research program in the face of disconfirmatory evidence. Thus, a Lakatosian perspective suggests that disconfirmation depends on the ability to visualize alternate hypotheses; visualization of alternatives depends, in turn, on how a subject or scientist mentally represents a problem.

Both Lakatos and Kuhn talk primarily about the role of competing scientific groups in providing these alternative explanations. My Eleusis study involved cooperation, not competition, within groups. But one could still see examples of processes amenable to a Lakatosian interpretation. One group was dominated by a subject who focused on suit to the exclusion of all other possibilities. On the final rule, which was "cards must alternate odd-even or red-black," he incorporated any evidence that appeared to contradict his "hard-core" idea that the rule had to involve just suit by making his guess more complex, so that the group's final guess looked like this: "In clubs, hearts, spades after first two cards in descending order as many cards after were allowed. In diamonds four descending cards then any card was allowed" (Gorman et al. 1984:76). Again, a Lakatosian perspective helps explain why this group proposed cards that falsified any simple suit rule, but did not abandon their focus on suit: they embarked on a "degenerating research program," altering the "auxiliary assumptions" of their suit "hard core" to fit new data until their final rule was almost incomprehensible. They were not able to conceive of an alternative explanation for their results.

FOUR

How the Possibility of Error Affects Falsification

One of the major criticisms that can be leveled at the experiments reported so far is that they are "unrealistic"—the tasks are not actual scientific problems and the subjects are not scientists (Brown 1989). To put it in psychological terms, they lack ecological validity, or what Berkowitz and Donnerstein (1982) disparagingly call "mundane realism." To have ecological validity, an experiment must be closely analogous to some real-world situation.

A similar criticism could be leveled at Popper: his views make sense in highly artificial situations, when one is testing a hypothesis like "all swans are white," but as tasks designed to simulate scientific reasoning become more complex, the effectiveness of falsification depends on other factors, such as the relationship between hypothesis and rule and the availability of domain-specific heuristics.

4.1. TWO TYPES OF VALIDITY

The power of the experimental method is that it allows us to abstract certain features from the noise of the "real world" and study them under controlled conditions (Houts and Gholson 1989). Berkowitz and Donnerstein argue that "artificiality is the strength and not the weakness of experiments" (1982:256). The higher the ecological validity, the more complex the situation, until it becomes impossible to be sure what variables are playing a causal role. Simple experiments, in contrast, can deliberately isolate variables of theoretical interest and assess their impact under circumstances that eliminate the noise due to other variables. These experiments can, in turn, suggest issues that ought to be studied in the field, in ecologically valid situations.

The experimental results reported earlier in this book have what Berkowitz and Donnerstein call external validity: they permit us to make theoretical generalizations about how and whether subjects employ falsification under specific conditions. For example, we can argue that disconfirmatory instructions will improve subjects' ability to solve a rule that is more general than the subjects' current hypotheses, provided the subjects represent the task in a way that suggests where to look for disconfirmatory information. Strictly speaking, generalizations of this sort are limited to situations and subject pools very similar to those used in the original experiment; additional experiments can be used to extend the range to different tasks and situations.

Steve Fuller (1989) has suggested that one could use externally valid experiments as a basis for developing philosophical norms. His concern is whether an experimental effect is reproducible outside of the laboratory, not whether the original experiment was representative in some mundanely realistic sense. Therefore, one could use experiments to help determine under what circumstances scientists ought to adopt a disconfirmatory heuristic. Conversely, if falsification turns out to be problematic even under ideal, externally valid conditions, it will have even less value in ecologically valid settings.

The radical implication of Fuller's view is that one ought to be able to derive testable conclusions from any philosophical approach that purports to have normative implications. At the time I began my next line of experiments, I had not read Fuller, of course, but he and I were struggling separately toward the same view. I wanted to explore the implications of various philosophical positions by adding variables to my simulations. At the same time, these variables might increase the ecological validity of my experiments. In particular, I became aware of some work at Bowling Green on error.

4.2. HOW THE POSSIBILITY OF ERROR AFFECTS FALSIFICATION

> In my laboratory, I find the laws of nature formally contradicted at every hour, but I explain this away by the assumption of experimental error. I know that this may cause me one day to explain away a fundamentally new phenomenon and to miss a great discovery. Such things have often happened in the history of science. Yet I shall continue to explain away my odd results, for if every anomaly observed in my laboratory were taken at its face value, research would instantly degenerate into a wild-goose chase after imaginary fundamental novelties. (Polanyi, quoted in Hon 1989:473)

Scientists work in an environment where error in the data is always possible; they learn not to discard a hypothesis simply because it is apparently contradicted by one or two experimental results. The possibility of error may be one reason why scientists often prefer a confirmatory heuristic. As Klayman and Ha noted, a confirmatory or positive test strategy "is actually a good all-purpose heuristic across a wide range of hypothesis-testing situations, including situations in which rules and feedback are probabilistic" (1987). When error is present, results are probabilistic, i.e., an experiment that should provide evidence for a hypothesis might appear to contradict it.

Error might even pose a problem for some of Popper's classic examples of falsification. Popper (1963) used "all swans are white" as an example of a hypothesis that would be decisively falsified by a single piece of evidence: a black swan. But as Holland, Holyoak, Nisbett, and Thagard (1986) note, upon encountering a black swan one could simply modify one of the subordinate or corollary hypotheses attached to one's general rule. For example, one could invoke error: perhaps the black swan was a rare mutation, a chance event that qualified as an exception to the rule, rather than as a falsification.

For Popper, Einstein was a scientist who always indicated how his hypotheses could be falsified. But, as noted above, Einstein never seriously considered the possibility that General Relativity would be falsified. Similarly, he dismissed an apparent falsification of Special Relativity, even though it was conducted by a distinguished physicist. Einstein argued that "whether there is an unsuspected systematic error or whether the foundations of relativity theory do not correspond with the facts one will be able to decide with certainty only if a great variety of observational material is at hand" (Holton 1973:235). As the physicist Paul Dirac noted, "It's most important to have a beautiful theory. And if the observations don't support it, don't be too distressed, but wait a bit and see if some error in the observations doesn't show up" (Judson 1984:43).

Another example of the role of error in scientific inference is cited by Beveridge:

> When I saw a demonstration of what is known as the Mules operation for the prevention of blowfly attack in sheep, I realised its significance and my imagination was fired by the great potentialities of Mules' discovery. I put up an experiment involving thousands of sheep and, without waiting for the results, persuaded colleagues working on the blowfly problem to carry out experiments elsewhere. When about a year later, the results became available, the sheep in my trial showed no benefit from the operation. The other trials, and all subsequent ones, showed that the operation conferred a very valuable degree of protection and no satisfactory explanation could be found for the failure of my experiment. It was fortunate that I had enough confidence in my judgment to prevail upon my colleagues to put up trials in other parts of the country, for if I had been more cautious and awaited my results they would probably have retarded the adoption of the operation for many years. (1957:34)

Again, Beveridge shows why scientists learn not to trust disconfirmatory results: they know "how difficult it often is to make an experiment come out correctly even when it is known how it ought to go" (Beveridge 1957:35).

Therefore, unlike subjects in most scientific problem-solving experiments, the working scientist has to be constantly aware that a single experimental result, even when methods are rigorous, may be due to chance. When a scientist encounters a potential falsification, he or she has to be careful—only reliable findings that have been consistently replicated should be taken seriously.

4.2.1. The Methodological Falsificationist's Solution

The examples above suggest how falsification might be preserved when error is possible: replicate. This is the methodological falsificationist's solution, as described by Lakatos: "One single observation may be the stray result of some trivial error: in order to reduce such risks, methodological falsificationists prescribe some safety control. The simplest such control is to repeat the experiment (it is a matter of convention how many times) . . ." (1978:24). In other words, methodological falsificationists consider that the possibility of error does not make disconfirmation much more difficult; it simply means that potentially disconfirmatory experiments such as Kaufmann's apparent disconfirmation of Special Relativity should be replicated.

From a psychological standpoint, the methodological falsificationist's assertion can be translated into an empirical claim. Is simple replication of disconfirmatory experiments sufficient to eliminate the effects of error on falsification? This claim can be explored experimentally, using tasks like Eleusis.

4.2.2. Hypothesis-Perseveration

A somewhat different prediction might be derived from Holland, Holyoak, Nisbett, and Thagard (1986) and Klayman and Ha (1987). The possibility of error might promote the use of a confirmatory or positive test heuristic and lead to a kind of hypothesis-perseveration. Kern (1982) conducted an experiment designed to determine whether error in the data could be used to protect subjects' hypotheses from disconfirmation. Her task was based loosely on the artificial universes used by Mynatt, Doherty, and Tweney (1977; 1978). Subjects were asked to determine which region on an imaginary planet could support life and which could not. Using a computer, they dropped probes containing plants called "tribbles"; the computer told them whether each tribble survived or not. From this data, subjects tried to place a boundary between the life and death zones on the planet's surface.

Kern compared two types of error: "system failure" and "measurement."[10] In the measurement error condition, subjects were told that the location of the probes on the planet's surface was known only within certain limits. In the system-failure error condition, subjects were told that on 25 percent of their trials the device that reported whether the tribble lived or died would fail, giving inaccurate feedback, i.e., if the tribble lived it would be classified as dead, and vice versa. It is this system-failure condition that comes the closest to the Einstein example above: Kaufmann's apparent disconfirmation should really have been a confirmation of Special Relativity.

Kern found that measurement error had no significant effect on subjects' performance, but system-failure error allowed subjects to "explain away" disconfirmatory data, and this adversely affected their ability to determine the actual boundary between life and death zones. In other words, system-failure error seems to have led to hypothesis-perseveration.

4.3. HOW THE POSSIBILITY OF ERROR AFFECTS THE "SEARCH FOR TRUTH"

Kern's study focused on the effect of actual measurement error and system-failure error. I decided to follow up her study by focusing on the mere possibility of system-failure error: does even the thought that there might be error cause hypothesis-perseveration?

To find out, I decided to add error to the same Eleusis design described in chapter 2. To simulate system-failure error, groups of four subjects were told that on anywhere from 0 to 20 percent of their experimental trials the feedback they received might be in error, i.e., a card that apparently fit the rule might *not* fit the rule and vice versa. Experimenters pretended to use random-number generators on calculators to determine which cards were errors. Groups were told to turn over any cards that they thought were errors. The instructions gave them an example, using the following sequence of cards:

10H	5C	9H	4H	KS	2H
	JC	3D			1S

They were told that to perceive the correct rule in this case, which was "alternating colors," they would have to turn over the 4H and the 1S (ace of spades). Groups were reminded that they could change their minds about error at any time, and turn cards face up again, or turn other cards over.

Three of the four Eleusis rules used in Gorman et al. (1984) were used in the present study; the second rule, "adjacent cards must be separated by a difference of less than three," was eliminated in the interest of time. Groups worked on their first task without possible-error instructions, to gain experience with Eleusis and each other; then, the possibility of error was introduced for the two final rules.

Groups were randomly assigned to one of the three strategy conditions used in the previous study. I expected that a disconfirmatory strategy would still prove markedly superior to confirmation because it would be so much easier for confirmatory groups to immunize their hypotheses against occasional disconfirmations by invoking error. Disconfirmatory groups, on the other hand, would produce so much negative evidence that they would be forced to discard false hypotheses, even under possible-error conditions.

No actual error was introduced. This study illustrates one of the advantages of experimental simulation: it allows us to separate factors that mingle in actual scientific practice. What is the effect of the mere possibility of system-failure error on scientific reasoning, as opposed to actual error?

4.4. RESULTS

My prediction concerning disconfirmatory instructions was itself disconfirmed. The possibility of error eliminated the effect of strategy noted in Gorman et al. (1984 and 3.3 above). Only five groups across all conditions solved Rule 1, before the possible-error instructions were introduced; only four—two disconfirmatory and one in each of the other conditions—solved Rule 2, and none solved the third rule.

Disconfirmatory groups did propose significantly more cards they predicted would be incorrect than did groups in other conditions, but even their proportion of predicted incorrects was low: about .12 as opposed to .03 in confirmatory and control. Disconfirmatory groups did not obtain significantly more incorrect cards than groups in other conditions did; on the average, disconfirmatory groups' cards were incorrect about 31 percent of the time, confirmatory's 27 percent, and control's 24 percent.

Except on Rule 1 (adjacent cards must be separated by a difference of one), a "warm-up" task on which performance in both studies was low, results under possible-error conditions contrast sharply with those obtained in the earlier, no-error study reported in chapter 2. Disconfirmatory groups solved Rules 2 (odd and even cards must alternate) and 3 (cards can alternate either red/black or odd/even) significantly more often under no-error than under possible-error conditions. In contrast, the performance of confirmatory groups did not differ significantly across the two studies (see Figure 4-1).

These results can be partially explained by the fact that, in the present study, disconfirmatory groups did not get significantly more cards wrong than confirmatory or control groups did, whereas in Gorman et al. disconfirmatory groups obtained a significantly higher proportion of incorrect cards and the proportion of incorrect cards played by groups in each condition on each rule was highly correlated with reaching the correct solution.

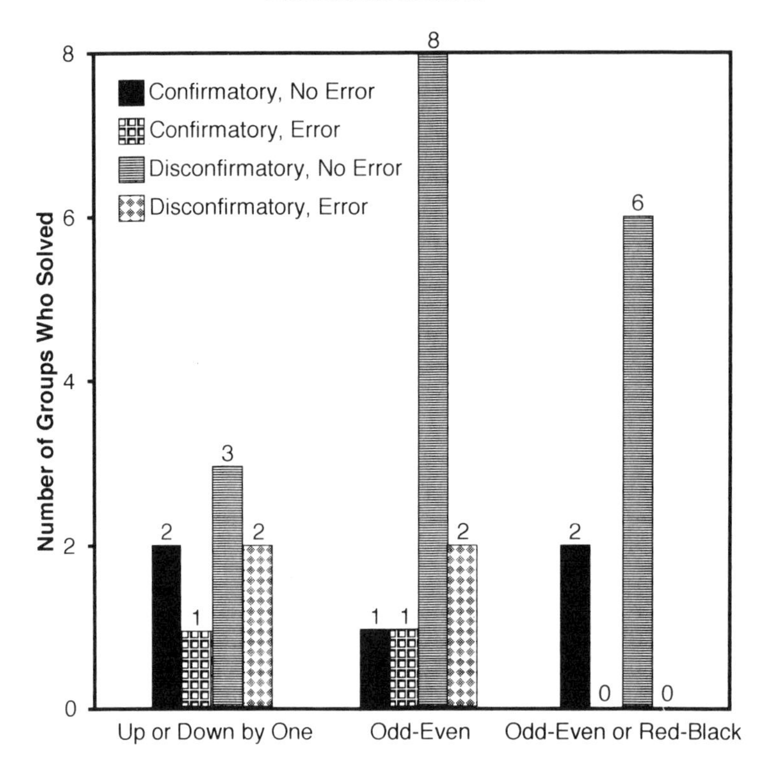

FIGURE 4-1 A COMPARISON OF CONFIRMATORY AND DISCONFIRMATORY GROUPS

4.4.1. ASSIGNING ERRORS

On Rule 2 (odds and evens alternate), only one disconfirmatory group had any cards still turned over at the end of their attempts to solve the rule, indicating that they thought there had actually been some error on the task. In contrast, half of the no-strategy (control) groups and all but two of the confirmatory groups left cards turned over. On rule 2, disconfirmatory groups did a better job of realizing there were no actual errors. On Rule 3, all but one confirmatory, two disconfirmatory, and two no-strategy groups left cards turned over at the end of the rule, indicating that most groups thought they encountered actual errors on this rule.

Overall, of the cards that remained flipped down at the end of each rule, 81 percent were disconfirmatory, i.e., they would have falsified the group's final guess if they had not been made errors. This result varied little across strategy conditions and rules and suggested that groups used errors to immunize their hypotheses against disconfirmation. Twelve of the groups that failed to solve Rule 2 and sixteen that failed on Rule 3 did so because they made into errors one or more cards that would have disconfirmed their hypotheses.

Consider the performance of a typical confirmatory group. By the time they were ready to stop working on Rule 2 (odd-even), their sequence looked like this:

1D	2S	3H	4C	11C	6S	7H	8C
				11D			
				5D			
				1S			

The underlined cards were considered errors—which means that the group thought the 11C should have been incorrect and the 5D correct after the 4C. They made these cards errors in an effort to preserve their final "ascending numbers" hypothesis.

This illustrates the typical effect of the error instructions: groups labeled as errors cards that didn't fit their hypotheses. No group solved the third rule, because it is hard to get cards wrong on that rule—either an alternating color or an odd-even pattern will produce endless right cards—and it is easy to assign errors to the few disconfirmatory results.

Consider the performance of a disconfirmatory group on Rule 3 (cards can alternate either odd-even or red-black). The cards up to this point alternated *both* odd-even and red-black except for the 11S, which they made an error:

1D	11S	2S	9D	2S
			9H	

Then they decided to deliberately play a card that should be wrong if the rule really was "cards alternate both odd-even and red-black." So they tried a 2D and a 5D, both of which were correct, disconfirming their hypothesis.

So they modified it: "cards must alternate in terms of color but all red cards are odd and all black cards are even." This meant the 2D had to be turned over. They immediately tried to disconfirm this rule with a 3S; when it was incorrect they turned it over also. Then they tried a 2C and a 9H, which were correct, and a 7D, which was incorrect, disconfirming their hypothesis. Instead of making the 7D another error, they decided to flip all their error cards back up and look for a simpler hypothesis. They tested whether odd-even was really an aspect of the rule by playing three even cards in a row—6C, 10D, and 12S, all correct. So they settled on a simple alternating color rule, flipping the 11S, 2D, and 3S back over again. Their final sequence looked like this, with error cards underlined:

1D	11S	2S	9D	2S	2D	5D	3S	2C	9H	6C	10D	12S
			9H						7D			

The error manipulation made it much harder for this group to test its hypotheses. In the end, it ignored three disconfirmatory cards. Without error, they would almost certainly have solved the rule, as they recognized both its alternating color and its odd-even components.

4.4.2. Groups That Distorted Error

Five groups represented the error instructions in a way that allowed them to preserve their hypotheses. Four of these replaced cards they had turned over with other cards that fit their hypothesis—even though they realized that this violated their error instructions. For example, one disconfirmatory group tried to make sense of the following sequence they had generated on Rule 2 (odd/even). (Again, underlining indicates cards that had been turned over):

1D	2C	3C	4H	9H	6S	13C	8D	9D	10H	11S	12H
				5S		13H					
				5S		1H					
				5D							
				1D							
				7H							

When the 9H was correct after the 4H, they had been working on a simple "ascending by ones" rule; therefore, they made the 9H an error—and made three attempts to follow it with a five. When all three were incorrect, they concluded that their "ascending numbers" hypothesis had been disconfirmed and searched for an alternative. But, with the exception of the 13C, the only cards they could get to work were ascending numbers. So they finally decided that they could *replace* an error card with another card; therefore, they turned the 9H and the 13C over and treated them as though they were a 5 and a 7 respectively. In other words, they treated error cards like blanks which could be filled by their choice of cards; this way of representing error allowed them to retain their hypothesis in the face of disconfirmatory evidence.

At this point, it is important to make a distinction between task representation and rule representation. An individual or group can have a particular way of representing the rule, and also a particular way of representing task constraints, such as what can and cannot constitute an error. Both types of mental representation affect choice of heuristic and performance.

On Rule 2, a confirmatory group which rarely deviated from its heuristic established a representation of the error manipulation, a feature of the task, that permitted them to retain their representation of the rule. Their cards at that point were as follows:

1D	2D	3D	4D	5D	10S	7D	8D	9D
13C					6S			
1C					6D			

One subject thought the 10S could be treated as though it were a 6D. Another subject disagreed, "because if there is error, it is as if the flipped card weren't there." A third subject agreed with both of them because "the six could also be error." In other words, at least one group member understood the error manipulation and realized that both the 10S and the 6D should be errors—but he didn't try to convince the others to turn the 6D over.

In a group, members don't necessarily share the same representations of either the task or the rule. Note that in this case the majority becomes convinced of one representation of error and the single member with a different view does not push his challenge beyond a certain point. The most radical misinterpretation of error occurred in a somewhat similar way, when one control group studied the following sequence of cards:

1D	2C	1H	2S	7D	8S	7H	8C
1C			2D				

One subject (whom we shall call L) said, pointing to correct cards one through four and cards seven through nine, "See—these are decreasing and increasing by one." Subject I pointed to the 7D and said, "But what explains this big jump here?"

L: "Error."

I: "How do we fix it?"

L responded by turning over the 7D. Five other "big jumps" were flipped over in the sequence.

Subject L gave his group an unusual representation of error: once a card was turned over, any other card could follow it. Here one subject's task representation became the whole group's representation, and made it possible for them to retain their hypothesis.

In conclusion, when they could see no other alternative, a few groups altered their representation of error in a way that fit their current representation of the rule. In the present study, the source of potential error was clearly defined in a way that prohibited such representations; even so, a few groups could not resist. In science, where the potential sources of error are numerous and not always known, it may be even easier to immunize a hypothesis by invoking error.

4.4.3. Groups That Assigned No Errors

There were twelve groups on Rule 2 and eight groups on Rule 3 that assigned no errors and still did not solve the rule. As noted earlier, on Rule 2 five of these groups were disconfirmatory. Four of these five settled on simple hypotheses involving a difference of one between adjacent cards and did not make sufficient attempts to disconfirm their hypotheses, in part because they were so concerned with making sure there was no error. In other words, the possibility of error made these disconfirmatory groups so preoccupied with error that it interfered with their attempts to falsify.

For example, in one group that guessed "cards must be increasing or decreasing by one," subjects spent several minutes arguing about error after they made their guess: one subject wanted to assign some errors, even though all the cards played fit their hypothesis. The argument was finally settled by referring to the instructions, which reminded them that there might be no error. In the amount of time it took this group to argue about error, they could have played a half-dozen disconfirmatory cards. As it was, they played only one card they predicted would be wrong: the final card, a 7S after a 7D. Even when groups assign no errors, the possibility of error can hinder their performance.

Another disconfirmatory group started by playing two aces and two kings, to make absolutely sure these cards had to be wrong after the opening ace. Then they began an ascending sequence, breaking it only occasionally. After the sequence reached the queen, they decided to test systematically to make sure there was no error by repeating the queen, then playing the king and repeating that three times. At this point, one subject said, "I put those kings down hoping to see error. If we play five kings in a row, then there's no error." Instead, the group went on to play two aces in a row, then took the sequence down to a nine.

This group correctly concluded that there was no error—but they spent so much time making sure of this that they did not try to disconfirm their hypothesis. Over 25 percent of the cards they played were replications of the card immediately preceding and all of these cards were incorrect. So the group did follow its suggested strategy by deliberately playing cards that should be incorrect, but most of these cards were replications of previous incorrect cards. In this case, even though a group recognized that there was no actual error, the mere possibility affected performance.

One of the groups in the confirmatory condition followed their strategy on Rule 2 by playing eight ascending cards in a row and concluded that the rule was "numerical sequence increasing by one." When one subject suggested replication as a heuristic—"You know, we should be trying these cards more than once"—another subject disagreed, saying, "the only way we can see error is to keep going." When their confirmatory heuristic produced a long string of correct cards, they correctly divined that there was no error but mistakenly identified their hypothesized rule with the target rule.

This group reinforces Klayman and Ha's (1987) observation that confirmation is a particularly useful heuristic in situations where results may be probabilistic. If a group can produce a long series of positive instances, then the possibility of error can be virtually eliminated, because any errors would break the pattern. Similarly, scientists often persist in their efforts to produce strings of confirmatory experiments. Tweney (1985) has documented how the physicist Michael Faraday struggled to obtain confirmations of electromagnetic induction, dismissing early disconfirmations as errors.

Once confirmation has eliminated the possibility of error, however, an effort must be made to disconfirm. To make sure that the "ascending numbers" pattern wasn't simply an aspect of a more general rule, the confirmatory group described above should have tried cards that were separated by differences greater than one. Faraday was not content merely to demonstrate that electricity could be induced from electromagnets; he also wanted to see if it could be generated from ordinary magnets, and if magnetism could be generated from a changing electrical current (Tweney 1985).

4.4.4. Disconfirmation Plus Replication

The two disconfirmatory groups that solved the second rule illustrate the importance of combining disconfirmation with replication when the possibility of error contaminates results and the hypothesized rule is embedded in the target rule. The first of these successful disconfirmatory groups began with a simple "ascending diamonds" sequence, ace through four. Then one subject said, "Try a diamond, but not a five—try a nine." It was correct. They tested to see if it was error by playing cards they thought would be right if the 9D were an error—a 3D, a 5C, a 5D. In other words, they realized that if the 9D were an error the next card in the sequence should follow the 4D. When neither the 3D or the 5C or the 5D worked, they felt fairly sure the 9D was not an error.

As of card 10, their sequence looked like this:

1D	2D	3D	4D	9D	10H	3H
				3D		
				5C		
				5D		
				11S		

One subject said, "It could just be any red card." They tried to disconfirm immediately, playing a 2C (correct) and a 4S (incorrect). They decided to guess that the rule was "any red card" and assigned errors to cards that appeared to disconfirm that rule.

Many groups would have stopped at this point, but this group continued. They played a 9D next, to replicate the situation in which they had first obtained incorrect red cards. Note how difficult it is to replicate on Eleusis: it is not enough to play the same card twice, one must play a card like the 9D in context with the same surrounding cards. Even this group's replication was not exact—at this point, the 9D came after a 4S, whereas earlier it had come after a 4D.

The 9D was correct. Now the group had re-created the situation in which red cards were wrong before. They combined replication with disconfirmation and got eight cards wrong in a row, seven of which were red. The maximum amount of error was 20 percent, so the "any red card" hypothesis was decisively disconfirmed.

The disconfirmatory instructions had also emphasized looking at incorrect cards, so one subject said, "Why don't we just look at these incorrect cards for a moment?" After a few minutes of study, another subject said, "I see something . . . odd-even." This rule fit all the correct and incorrect cards, including the potential errors, so they turned up all their errors and made their final guess after proposing a few more cards to check it.

Note that we cannot completely explain the actual moment of discovery in this group. Kuhn would rely on Gestalt psychology to indicate why one subject's perception shifted from seeing reds and blacks to seeing odds and evens. Perhaps this process resembles the kinds of paradigm-shifts that occur in science. What we can say is that eliminating alternative hypotheses and focusing on incorrect cards were necessary, if not sufficient, conditions for perceiving the correct rule.

So, this group carefully followed its suggested heuristic, seeking and studying incorrect cards. They also replicated situations in which they had uncovered disconfirmatory information. The best higher-order heuristic under possible-error conditions on tasks of this sort is a combination of disconfirmation and replication.

A second successful disconfirmatory group and the one successful confirmatory group also followed this heuristic. The disconfirmatory group began with a sequence that looked like this:

1D 2C 7H
13D
13S

They flipped the 13S over to make the cards follow an alternating red-black pattern. But they replicated their major disconfirmatory instance by playing the 13C after the 7H, to see if a black king would still be wrong after a red card. It was wrong again, and so was an 11C, disconfirming the alternating colors hypothesis. They proposed, tested, and eliminated two more hypotheses before one subject saw the odd-even pattern. Half of the cards they played were incorrect.

In response to the question "What was your suggested strategy" on the final questionnaire, three of the four members of the only successful confirmatory group specifically mentioned trying cards "that probably won't work" and "testing all extremes." They also replicated potentially disconfirmatory instances. They, too, began with an alternating color pattern and encountered a black king that did not fit it: as one subject said, "Alternate colors is one of the patterns, but there must be some other restriction because the black king didn't work." They replicated, playing several black kings, which were wrong where they should have been right; at first they labeled these anomalous cards errors, but repeated replications made it clear that their color hypothesis had been disconfirmed. Shortly afterwards, one subject saw the odd-even rule; they tested it by playing four odd cards in a row and two even cards in a row. Ignoring their strategy instructions and pursuing a combination of disconfirmation and replication was the key to their success.

The one control group that solved Rule 2 illustrates the value of Platt's strong-inference heuristic, combined with replication. At card 6, their pattern looked like this:

1D 2D 3S 4D 3S 2H

They considered three hypotheses: alternating colors, up and down by ones, and alternating odds and evens. (Note that the alternating colors hypothesis would have required them to make the 2D an error.) They decided to test these hypotheses by playing a 6S next, after the 2H. If it were incorrect, then the alternating-colors hypothesis would be disconfirmed, unless the 6S were an error. If it were correct, then the up-or-down-by-one and the odd-even hypotheses would be disconfirmed, unless the 6S were an error. This is Platt's heuristic: come up with two or more competing explanations for a result, then design a test to eliminate at least one.

The 6S was wrong, apparently disconfirming the alternating-colors pattern. But was it an error? To make sure, they immediately replicated the situation they had just tested—by first playing a 3H and then a 7S. The first one was right and the second one was wrong, which cast further doubt on the alternating-colors hypothesis.

Both of the other hypotheses were still intact. How to eliminate one? One subject suggested, "Try an even card that isn't a four." If it were right, then the up-or-down-by-one pattern would be disconfirmed; if it were wrong, then the odd-even pattern would be disconfirmed. They tried an 8C, which was correct. Another subject pointed out that "that could be an error"—so they replicated the situation with an 11C and a 2C, both of which were correct. They had eliminated all hypotheses but odd-even. As one subject said, "We could keep going to find some other rule but something as simple as odd-even seems to hold."

Tweney et al. (1980:119) found that individuals working on the 2-4-6 task had trouble employing Platt's strong-inference heuristic: "subjects preferred to evaluate several pieces of data against a single hypothesis, rather than one datum against several hypotheses" (see 3.4 above). Groups, however, could keep track of several hypotheses at once, because each individual can be working on a different theory. For the group considered above, competing hypotheses helped them focus on cards that might constitute critical experiments. Under error conditions, no single experiment is sufficient to distinguish between competing hypotheses, of course, so they had to replicate their critical cards and their surrounding context. Alternative hypotheses suggest directions to look for possible disconfirmations and experiments that should be replicated.

4.5. CONCLUSIONS

The methodological falsificationist's advice appears difficult to follow when there is a possibility of system-failure error, even when there is no actual error. Instead, groups were so determined to preserve their hypotheses that some even altered the way they represented error, even though it was clear that they had originally understood the instructions. Even in cases where a group assigned no errors, attempts to replicate often interfered with disconfirmation. Successful groups combined replication with disconfirmation.

One referee on my first attempt to publish these results pointed out that one explanation for the difference between the possible-error and no-error Eleusis studies is that the former results were themselves due to error. For example, subject-pools for the two studies were quite different: the original study without error was run at a state university with a diverse student body (see Gorman et al. 1984), whereas the later study was run at an engineering school whose students were all well-versed in science and mathematics (see Gorman 1986). Perhaps the difference in results really only reflected a difference in subjects.

I was able to satisfy the referee by pointing out that results on the first, "warm-up" rule in both studies were very similar, suggesting that the two subject-pools were relatively equal in terms of their ability to solve Eleusis rules. The study was eventually published (Gorman 1986). But this referee had alerted me to the very problem I was studying. If I believed my own results, I ought to replicate.

However, I was not inclined to follow the methodological falsificationist's heuristic: a straightforward replication would be difficult to publish. I was in a situation similar to the one I had been in after my first Eleusis experiment, and I repeated the heuristic I had developed on that occasion: I went from a study using Eleusis and groups to one using the 2-4-6 task and individuals. My first 2-4-6 study had replicated the results obtained with my first Eleusis study, but had extended them to individuals working on a new task and rule. I expected that adding the possibility of error to the 2-4-6 task would replicate and extend the Eleusis results reported in this chapter.

4.6. ADDING POSSIBLE ERROR TO THE 2-4-6 TASK

Would the Eleusis results generalize to individuals working on the traditional (Wason 1960) version of the 2-4-6 task? If so, the possibility of error should encourage subjects to retain their hypotheses in the face of potential falsifications, reducing their ability to solve the rule as compared with a control group. However, if the methodological falsificationists were right, subjects working under possible-error conditions should replicate triples significantly more often than would subjects in the no-error condition, and these replications should enable them to solve Wason's rule as often as would subjects working under no-error conditions.

Note that "solution" in this case is defined as finding the target rule, which is embedded in most of the rules subjects are likely to hypothesize at the outset. The relationship between hypotheses and target is the same on the Eleusis rules used in the last chapter. Therefore, results ought to be similar, even though this study focused on individuals rather than groups. This shift was made to make sure that results from groups generalized to individuals.

4.6.1. Procedures

Eleven volunteers from an introductory psychology class at Michigan Tech were randomly assigned to a possible-error condition and twelve were assigned to a control, no-error condition. Subjects were run about six at a time, but no interaction between individuals was permitted. After they were told how the 2-4-6 task worked, subjects in the possible-error condition were given an additional sheet of instructions as similar as possible to those used in section 4.3 (see Gorman 1986), emphasizing that feedback on anywhere from 0 to 20 percent of subject's triples could be erroneous, i.e., if a triple were correct it would be classified as incorrect and vice versa. Error could be determined by random-number generators on calculators. To indicate where errors occurred, subjects in the Eleusis study turned cards over; subjects in the present experiment were instead told to put asterisks by triples they thought were erroneously classified.

Wason's (1960) "ascending in order of magnitude" rule was used, and his task was run in a manner identical with my previous studies (see chapter 3). No actual error was introduced.

4.6.2. Results

Although the possibility of error had a significant effect on groups' abilities to solve Eleusis rules, individuals' abilities to solve the 2-4-6 task were not affected: eight of eleven control subjects and nine of twelve possible-error subjects solved the rule. Subjects in the possible-error condition did have to propose an average of twenty triples to achieve this level of performance, whereas control subjects needed an average of only thirteen. But the former did not replicate significantly more than the latter: only three subjects in the possible-error condition proposed the same triple more than once, as opposed to one subject in the control condition. Furthermore, only three subjects in the possible-error condition asterisked any triples. Apparently, subjects did not need to follow the methodological falsificationist's advice in order to determine that there were no errors.

The subject who assigned the most errors began the "ascending numbers" rule with triples like 3,6,9 and 4,8,16; his first hypothesis was "the (second and third) numbers are multiples of the first number." Then he followed a disconfirmatory heuristic, proposing triples like 21,31,41 and 5,7,11, which should have been incorrect if his hypothesis were correct. When these triples were correct, he marked them with asterisks and turned to a confirmatory heuristic, proposing triples like 5,25,100, which followed his initial hypothesis. By the end, he had marked six triples with asterisks. As he said on a final questionnaire, in response to a question about what strategy he had followed, "I marked the ones that didn't correspond with an asterisk. . . . I figured, due to the error, that a few of these yeses were nos." This subject showed that the possibility of error could be used to preserve a hypothesis against disconfirmation. But only three subjects actually assigned any errors.

The subject above failed to replicate any of his triples. In contrast, let us consider the responses of a subject who followed the methodological falsificationist's advice. This individual proposed more triples than any other subject (thirty) and also replicated more often (five times). He began with the triples 1,3,4, 1,2,3, 3,6,9 and 0,1,1 and then replicated all four. As he said on his final questionnaire, in response to a question about error, "At first I thought there was a definite 20 percent error, but reading the directions over, I found that it could be 0–20 percent. I now assume there was 0 percent error!"

His first hypothesis on the "ascending numbers" rule was "the sum of the first two numbers must equal the third." Later, he added "no two numbers can be the same," to explain why 0,1,1 was incorrect. But when he switched to a disconfirmatory heuristic, proposing 1,2,4, he found that triple was correct. He asterisked it, but also charged his hypothesis: "numbers have to be in increasing order." He tested this hypothesis with confirmatory triples like 0,1,100 and disconfirmatory triples like 3,2,1. Finally, he replicated 1,2,4, establishing that it was not an error. This illustrates how the methodological falsificationist's heuristic can lead a subject to realize that there is no error and to divine the correct rule under possible-error conditions.

Klahr and Dunbar (1988) noted similar behavior in an experiment that inadvertently included some possibility of error from the standpoint of at least one subject. This subject was trying to learn the function of a "repeat" key on a device called a Big Trak. When a result violated his expectations, he found it "strange" and repeated the trial. It is not clear that he was worried about system-failure error; instead, he probably just wanted to make sure he had conducted the previous trial and encoded its result properly. At any rate, replication was sufficient to convince him that this disconfirmatory result was accurate, and he had to change his hypothesis about how the repeat key worked. Immediately after his replication, he identified the device's correct function (see Klahr and Dunbar 1988:12–15). For this subject, the methodological falsificationist's heuristic was sufficient.

But this heuristic does exact an obvious price: replication takes time and uses resources. The most "efficient" possible-error subject solved the rule in the least number of triples (eleven) without replicating or assigning any errors. His first hypothesis was that each correct triple was some simpler triple doubled: 8,10,12 was correct because it was 4,5,6 doubled, 10,14,18 was correct because it was 5,7,9 doubled, etc. Basically, this vague rule could be made to fit all the correct triples he obtained, so he assigned no errors. On his questionnaire he said, "order appeared to be the key so I replaced all hypotheses about numerical relationships with a hypothesis of order of magnitude." In other words, he appears to have used some kind of aesthetic criterion, such as "simplicity."

He also said he "tested some cases more than once for eliminating the error," even though there were no exact replications among his triples. Instead, he appears to have repeated earlier patterns that worked, proposing 8,10,12, 10,14,18, and −6,−4,−2 to confirm that the "doubling" pattern was not simply due to error.

This represents a new heuristic: replication-plus-extension. Replication involves conducting exactly the same experiment more than once; replication-plus-extension involves conducting an experiment that is similar to one conducted before but which contains one or two novel features.

Consider, for example, the relationship between the possible-error experiment reported in this section and the Eleusis error experiment reported in 4.3 and 4.4. The former is a replication-plus-extension of the latter; the possible-error manipulation remains the same, but the task is different and subjects are run individually instead of in groups. The replication failed. Is it because the first result was due to chance, or because the "extensions" altered the situation too greatly? Further experiments were required to find out, and will be reported in the next chapter.

But first, let us conclude our discussion of how the possibility of error affects performance on Wason's 2-4-6 task. Even the possible-error subject who proposed the fewest triples checked for error, using a replication-plus-extension heuristic. This explains why control subjects needed fewer triples to solve the rule. Consider the control subject who proposed the fewest triples (eight). The triple 2,4,6 inspired his first hypothesis: "the first two numbers are added to find the third." He proposed 8,10,18, a confirmatory triple, to test it. He then came up with an alternate hypothesis: "all the numbers in the triple have to be even." He then followed Platt's strong-inference heuristic by proposing 2,4,8, a triple that disconfirmed the adding rule but confirmed the "all evens." He then proposed 1,4,5, which disconfirmed the "all evens" hypothesis. Notice that he felt no need to propose any replications or replications-plus-extensions of these disconfirmatory triples; in the absence of any error, he could design "crucial experiments" using a single triple. As he said on his final questionnaire, "When both [of these hypotheses] tested out as unimportant, I noticed that as long as the numbers were in ascending order they were correct."

Like so many subjects in these possible-error experiments, I now had two results which should have been the same, according to my hypothesis, yet one was positive and the other negative. Perhaps replication-plus-extension was easier on the 2-4-6 task than in Eleusis. Or perhaps the experimenter studying error was himself a victim of error: either the Eleusis or the 2-4-6 result might be due to chance. (From the standpoint of the subjects, this would have been a fitting revenge for the work I put them through.)

The only way to find out was to replicate or replicate-plus-extend. But, as we shall see in the next chapter, there were costs associated with this decision—costs I was not sure I wanted to bear.

FIVE

Pursuing the Possibility of Error

When I added the possibility of error to the 2-4-6 task, I followed a heuristic I had pursued successfully in my first two publications: when a manipulation works on Eleusis, try it on the 2-4-6 task. However, results on the 2-4-6 task appeared to disconfirm results on Eleusis: the introduction of the possibility-of-error factor had a strong effect on subjects' performance on Eleusis, and virtually no effect on the 2-4-6 task. What was I to conclude? That the possibility of error had (1) a minimal or (2) a major impact on tasks that simulated scientific reasoning?

There was no guarantee that further studies would clarify this paradox—I might invest years in chasing an elusive possible-error effect that only I was interested in. Any good scientist or research team can discover or generate a virtually infinite number of new ideas; only a few of these can be actively pursued to the point where they will achieve the sort of justification traditional philosophers talk about (see 3.7 for a discussion of the context of pursuit). How do scientists decide which ideas are worth a major investment of time and energy? At this point, I was facing this sort of decision myself: I could mount a major program of research to determine why the possible-error effect occurred in one study but not in another, or I could simply abandon the problem as unlikely to yield promising results quickly enough.

5.1. THE SOCIOLOGY OF PURSUIT

What constitutes a "promising result"? Clearly, I was not primarily concerned with results that yielded valuable information about the possibility of error—almost any well-conducted experiment would yield some valuable information in this regard, even if it only made the picture more complicated. No, my concern was whether and how quickly I would achieve publishable results, especially if quite a few studies led up blind alleys. After all, I had really invented this possible-error research problem; it didn't follow up on some promising line of research already recognized by the major journals, and therefore all but the most clear-cut and spectacular results might be difficult to publish. I was encouraged by my success with the *British Journal of Psychology* and the *Quarterly Journal of Experimental Psychology*, but I couldn't keep coming back to the same journals forever. Tenure and promotion—or the alternative—loomed on the horizon. Should I abandon this program of research?

Had I a laboratory and teams of graduate students, I could have carried on several research programs simultaneously—in which case, one can gamble on a higher-risk venture that is sustained by success in more traditional areas. But I had to rely on undergraduate volunteers as research assistants and borrow space wherever I could find it. Therefore, I could really maintain only one experimental research program at a time.

Latour and Woolgar (1986), in their famous ethnographic study of the Salk Institute for Biological Studies, explored how scientific knowledge resulted from social negotiations. In the course of this study, they made observations relevant to the sociology of pursuit decisions. Before considering the relationship between their work and pursuit, we need to review their methods and assumptions, in part because they represent such a stark contrast to the experimental work that has been the subject of most of this book.

5.1.1. Following Scientists Around

Latour and Woolgar's method was basically to "follow the scientists around," being careful not to "go native" by privileging the language or customs of the scientists:

> We take the apparent superiority of the members of our laboratory in technical matters to be insignificant, in the sense that we do not regard prior cognition (or in the case of an ex-participant, prior socialisation) as a necessary prerequisite for understanding scientists' work. This is similar to an anthropologist's refusal to bow before the knowledge of a primitive sorcerer. For us, the dangers of "going native" outweigh the possible advantages of ease of access and rapid establishment of rapport with participants. Scientists in our laboratory constitute a tribe whose daily manipulation and production of objects is in danger of being misunderstood, if accorded the high status with which its outputs are sometimes greeted by the outside world. (Latour and Woolgar 1986:29)

It is hard to imagine an approach more different from the philosophically oriented experiments discussed so far in this book. But if one keeps the distinction between ecological and external validity in mind, a reconciliation becomes possible. To review, an ecologically valid study is one that is highly realistic; an externally valid study, in contrast, is directed primarily at theoretical concerns and therefore can be quite artificial. Latour and Woolgar followed real scientists around their laboratories, observing their daily activities in detail; therefore, their research potentially has high ecological validity.

Note the emphasis on" potentially" here. It is possible to study scientists in a realistic setting using improper or idiosyncratic methods, which would reduce or eliminate one's claim to ecological validity, just as a poorly designed experiment lacks external validity. This is the issue of internal validity, which has to do with whether the chosen methodology is executed appropriately.

Latour and Woolgar claim their approach is atheoretical, though they obviously wish to make generalizations about the behavior of scientists. The goal of experimental work, in contrast, is to generate and/or test theories about specific aspects of science; therefore, such studies can be conducted in artificial settings with nonscientists. If these experiments are conducted properly, their results will have external validity, even if it is not clear how the variables under study will interact with other variables in a realistic scientific setting.

5.1.2. Investment Heuristics

Latour and Woolgar maintain that "It is at best misleading to argue that scientists are engaged, on the one hand, in the rational production of hard science and, on the other, in political calculation of assets and investments. On the contrary, they are strategists, choosing the most opportune moment, engaging in potentially fruitful collaborations, evaluating and grasping opportunities, and rushing to credited information" (1986:213). They point out that the scientists they studied "related how they . . . drifted until they found an instrument, a method, a collaborator or an idea that worked" (p. 212).

While they do not use the term, Latour and Woolgar are discussing how scientists make "pursuit" decisions. To combine ethnographic and experimental generalizations, an "idea that works" can be a hypothesis that yields confirmatory results. From the sociologists' standpoint, one uses this confirmation heuristic not merely to "search for truth" but to gain credit in the scientific community. If a young scientist is trying to decide where to invest his or her efforts, confirmation is a useful heuristic—follow a research program that produces positive results. Giere (1988) discusses how young physicists made decisions to invest in particular models because those models made predictions which could be confirmed. This strategy should produce more publications; Mahoney (1976) has discussed how scientific journals, at least in psychology, are biased toward confirmation (see 7.2.).

Disconfirmation of others' ideas can be a successful investment heuristic as well. As Mitroff (1974) points out, scientists are often motivated to disprove another's hypothesis; this sort of falsification advances the career of the one doing the disproving at the expense of the one disproved.

So, an optimal investment strategy for a scientist might be one that combines confirmation of one's own hypothesis with disconfirmation of a rival's—which fits in nicely with Lakatos's ideas concerning competing research programs. Some of Latour and Woolgar's examples may be amenable to this sort of analysis; for example, Guillemin's pursuit of a reliable bioassay for TRF (thyrotropin releasing hormone) involved obtaining rapid confirmations that also "shot down a number of existing claims" (Latour and Woolgar 1986:213).

In establishing a strong effect for disconfirmatory instructions, I had followed this investment heuristic: I confirmed the existence of such an effect in two studies and, in so doing, disconfirmed the consensus in the field, which was that disconfirmatory instructions didn't work. I was following only part of this investment heuristic in my possible-error research: I was trying to confirm my own effect, but there was no earlier research to disconfirm. So further attempts to replicate or to extend this possible-error effect were already on shakier ground.

In the investment terms outlined above, it might have made more sense for me to pursue a different line of research. I wondered if I had stumbled on a research paradigm that would lead to results which might be difficult to publish—might, in fact, never be published. I was drifting farther away from my invisible college and was in danger of becoming a one-man cottage industry. But this very paradox was motivating: I was determined to find out why I got an effect in one study and not in another.

How to proceed? Well, I knew better than to conduct a straightforward replication of my earlier work; I knew by now that journals wouldn't publish it, and, as a scientist quoted in Mulkay pointed out, "It's both boring, uninteresting and unpublishable, *just* to repeat (an experiment). It's really (worthwhile) only if you can add something. If you can find out *why* something has happened, instead of just saying 'this happened' " (1988: 92).

My long chain of error experiments was motivated both by the need to make sure the possible-error effect was a stable phenomenon and by the need to discover why it occurred. Like the scientists interviewed by Mulkay, I never conducted an exact replication; instead, I tried to replicate an effect under conditions that varied in specific ways from its original demonstration. In short, I followed the replication-plus-extension heuristic that my own research had found to be so valuable on the 2-4-6 task.

Of course, each replication-plus-extension introduced new potential confounds. For example, when I switched from Eleusis to the 2-4-6 task I not only changed tasks, I went from groups to individuals. My earlier experiments suggested that the effect of disconfirmatory instructions generalized well across the two tasks and with both groups and individuals, but there was no guarantee that the possibility of error would operate the same way. I had at least one study which suggested that this possibility of error had an externally valid effect. The remaining question had to do with generalizability: was this effect limited to a very particular set of circumstances, and, if so, what were they?

5.2. ADDING TWO DIMENSIONS TO THE 2-4-6 TASK

One possible explanation for the different possible-error results on Eleusis and the 2-4-6 task is that the former is more complex than the other; whereas rules in the 2-4-6 task involve numbers only, rules in Eleusis could also involve color or suit. Coupling error with these additional dimensions might "cognitively overload" the subjects to the point where adding the possibility of error would have a disruptive effect. Klahr and Dunbar (1988) found that subjects sometimes reacted as though there might be error on a complex, multidimensional task (see sections 4.6.2 and 9.2.2).

To make the 2-4-6 task more like Eleusis, two dimensions were added: color and letter. Subjects could write their triples in either red or black or in any combination of the two, and every triple was followed by one of the 26 letters of the alphabet.

I decided to use one of the Eleusis rules, to make the tasks even more similar. The Eleusis rule on which the clearest difference between possible-error and no-error conditions occurred was "odd and even cards must alternate." So I developed an analogous 2-4-6 rule: "numbers in the triple must alternate odd and even."

One additional feature was added to the 2-4-6 possible-error instructions to make them more similar to those used in the Eleusis study. Eleusis subjects turned over cards they thought were errors, whereas subjects in the study described in 4.6.1 only asterisked triples they thought were errors. This meant that the Eleusis subjects could, in fact, ignore potential errors whereas the 2-4-6 subjects could still see them. To eliminate this difference, 2-4-6 subjects in this study were given adhesive strips to cover triples they thought were errors. They could remove or rearrange these strips at any time to reflect changes in their perceptions of error.

It was expected that these procedural changes would make the performance of subjects under possible-error conditions virtually identical on the two tasks; i.e., subjects in the possible-error condition would solve the rule significantly less often than control subjects would, despite making more efforts to replicate and/or replicate-plus-extend.

5.2.1. Results

Seventeen student volunteers from Michigan Technological University were randomly assigned to a possible-error condition and twenty-one were assigned to a control, no-error condition; two of the seventeen possible-error subjects and four of the twenty-one no-error subjects solved the rule. While fewer possible-error than no-error subjects solved the rule, the difference was too small to be statistically significant and could easily have been due to chance. Possible-error subjects did, however, propose significantly more triples and replicate significantly more often (see Gorman 1989c. for details of statistical tests). Subjects in the possible-error condition proposed an average of 26.4 triples, as opposed to an average of 18.5 for subjects in the control condition. Nine subjects in the possible-error condition replicated triples two or more times; the eight other subjects did not replicate. Only four subjects in the control condition replicated a triple once; the other thirteen did not replicate any triples. Six of the seventeen subjects in the possible-error condition assigned errors, twice the proportion as in the earlier possible-error experiment on the traditional version of the 2-4-6 task.

To assess the effect of the two extra dimensions, I kept track of one additional variable on this version of the 2-4-6 task. I wanted to measure the extent to which subjects followed a heuristic I labeled "holding constant," in which one variable would be systematically varied while all others were kept at a constant value. On this multidimensional task, subjects had the option of holding two dimensions constant while varying a third. For example, to find out whether color really makes a difference in the rule, a subject could repeat a triple like 2,4,6,L several times, varying the color but holding both number and letter constant.

Bruner, Goodnow, and Austin (1956), who studied concept formation on multidimensional tasks, called the same heuristic "conservative focusing" and contrasted it to "focused gambling," in which a subject changes more than one attribute or dimension at a time. Focused gambling offers the potential for quicker solution of the rule in situations where the dimensions interact in an unexpected fashion. For example, supposing our hypothetical subject above breaks the sequence of 2,4,6,L triples, all of which represent various combinations of color, by trying something completely different: say, 9,4,7,E in a novel color combination. If this triple is correct, it may allow the subject to form a new hypothesis, especially if it reveals that the target rule lies outside of the subject's hypothesized rule. But the subject can't be sure *why* the triple is correct; to test the interaction between dimensions in this new hypothesis systematically, the subject will have to return to holding constant. Once again, the relative value of heuristics depends on the subject's mental representations and whether she or he is in the pursuit or testing phase of the task.

I could not find a way to operationalize focused gambling, but I could measure holding constant by simply checking the proportion of triples on which two dimensions had the same value. Overall, subjects held 62 percent of their triples constant; there were no differences between possible-error and no-error conditions on this measure.

5.2.2. Tentative Conclusions

Adding two dimensions to Wason's traditional 2-4-6 task encouraged possible-error subjects to propose more triples and to replicate more often than control subjects. But there were no significant differences in performance; in fact, the added dimensions made the task so difficult that only six out of thirty-eight subjects solved it.

Subjects in both conditions successfully followed a "holding constant" heuristic: all but four eventually realized that the rule involved only the number dimension; three of these thought the rule involved both number and letter and one thought it involved all three. While this heuristic was effective at identifying which dimensions were really critical, it apparently interfered with subjects' tests of their final hypotheses. Ten possible-error and nine control subjects thought the rule involved a difference of one between adjacent numbers in the triple. Similarly, when the possibility of error was added to Eleusis, many of the groups who successfully used replication to eliminate the possibility of error settled on final hypotheses involving a difference of one between adjacent cards.

The fact that possible-error subjects replicated more often than did control subjects, but did not employ holding constant with any greater frequency, highlights the similarities and differences between these two heuristics. Holding constant involves replication of two dimensions in order to assess the role of a third; replication-plus-extension involves making a later triple very similar to an earlier one in order to determine the role of error. Apparently both can interfere with subjects' ability to employ a disconfirmatory heuristic.

5.2.3. Piloting and Pursuit

At this point I moved back to the pursuit phase of research; instead of immediately launching into a full-fledged study, I would pilot several methods for improving subjects' performance to the point where a possible-error/no-error difference might appear. Behind virtually every published experiment is an extensive series of such pilot studies, where one tinkers with procedures and variables to find a promising combination. Such tinkering is never referred to in a published report, except perhaps as a footnote, but the most significant discoveries often occur when piloting.

My Ph.D. advisor years before had taught me a heuristic: pilot until you are certain what the results will be, then run a full-fledged experiment. Note the emphasis on confirmation as an important pursuit heuristic, but in a specialized sense. One pilots to pursue an effect, then—once it has been demonstrated—one conducts an experiment that will confirm it for the field. The effect or result itself may disconfirm a theory. So confirmatory and disconfirmatory heuristics may be applied to effects or to theories; a heuristic that confirms an effect may disconfirm a theory, and vice versa. Klayman and Ha (1987) were making a somewhat similar point when they argued that a positive test strategy may be confirmatory or disconfirmatory depending on the relationship between hypothesis and target rule; however, they did not discuss the role of positive and negative test heuristics in the pursuit or pilot phase of inquiry and therefore missed the fact that one can confirm an effect which disconfirms a theory. For example, when Faraday confirmed that a polarized light ray was slightly rotated if it passed through a magnetic field he also disconfirmed current notions about the ether (Tweney 1984).

5.2.4. Multidimensional Disconfirmation

To illustrate these pursuit processes, I want to continue unpacking the black boxes my published reports may have created. Perhaps if subjects on the multidimensional 2-4-6 task were given disconfirmatory instructions, their performance would improve as dramatically as did that of Eleusis groups, every one of which solved the odd-even rule in a disconfirmatory condition. I ran a pilot study to find out whether disconfirmatory instructions would produce a difference between possible-error and no-error conditions on this multidimensional 2-4-6 task. Ten subjects were run, five in a possible-error condition and five in a control condition; all ten were given disconfirmatory instructions of the sort similar to those used in previous studies. Overall, these disconfirmatory instructions encouraged subjects to propose both more triples and a higher proportion of incorrect triples than were proposed by subjects in the previous experiment with the multidimensional 2-4-6 task. Disconfirmatory subjects proposed an average of 32 triples and got almost half (45 percent) wrong across both conditions; in the larger experiment, without strategy instructions, subjects proposed an average of 22 triples across conditions and got about 30 percent wrong. The net effect was to double the percentage of solvers, from roughly 15 percent in the experiment without disconfirmatory instructions to roughly 30 percent in the pilot study with such instructions.

I had found a way to slightly increase performance on this multidimensional task and, not incidentally, I had conducted a potential replication-plus-extension of the positive effect for disconfirmatory instructions. Note that I had also applied the heuristic of holding constant: I had kept everything but strategy instructions the same between this study and the previous one. Of course, this was only a pilot study with a very small sample; to verify these tentative generalizations I would have had to conduct a full-fledged experiment.

There were, however, almost no differences between possible-error and control conditions: two of the five possible-error subjects solved the odd-even rule, as did one control subject. Four possible-error subjects replicated at least one triple; two control subjects did the same. Proportions of incorrect triples and triples held constant were very similar. Possible-error subjects did propose an average of 37 triples, as opposed to 28 in the control condition; this result suggests that possible-error subjects were engaging in more replication-plus-extension.

I had expected to see a difference strong enough to emerge even with only a few subjects when disconfirmatory instructions were used. Instead, results resembled that of the previous 2-4-6 possible-error studies: possible-error subjects made effective use of replication-plus-extension to check for error. Now I was faced with a classic pursuit decision. Should I invest my limited time and resources in making this a more thorough replication, or should I pursue other possible explanations for the lack of a possible-error effect on the multidimensional 2-4-6 task? My previous research and training had given me a mental representation of what constituted a successful result: a difference large enough to see with a small number of subjects. Clearly, disconfirmatory instructions were not going to produce this sort of difference on this task.

A paper by D. D. Tukey gave me a theoretical justification for abandoning strategy instructions. He conducted a series of experiments on the 2-4-6 task in which he investigated the value of philosophies other than Popper's for explaining subjects' responses—but he deliberately made no effort to manipulate their behavior via strategy instructions. As he said, "Several recent studies on Wason's 2-4-6 task have tried to modify subjects' methods via manipulating task instructions (e.g., Tweney et al. 1980; Gorman and Gorman 1984). Rather than legislating what subjects should be doing in this task and then trying to evoke that behaviour, the present research suggests that psychologists learn from subjects which methods or strategies are easily and/or beneficially explored" (Tukey 1986:30). I resented the pejorative use of the term *legislate;* Tukey seemed not to understand that if one wanted to make externally valid claims about the effectiveness of various strategies, one had to give subjects instructions encouraging (not "legislating") their use. But his advice was good if one's goal were to use experiments as a means of generating hypotheses. He was able to see significant evidence of subjects operating in a Lakatosian rather than a Popperian fashion on the 2-4-6 task, engaging in degenerating research programs and the like; I was independently finding similar evidence (see 3.9.1).

In the possible-error studies, I now had evidence that disconfirmatory instructions did not differentially effect possible-error and no-error conditions, despite my earlier Eleusis results. I could have attempted to establish this lack of effect with a more thorough study, but I was (a) leery of trying to publish a negative result and (b) more interested in the possibility-of-error paradox. Therefore, even though Tukey seemed somewhat critical of my work, I saw an opportunity to follow one of Latour's (1987) investment heuristics and "recruit him as an ally" by citing Tukey in a future publication as a justification for *not* combining strategy instructions with the possible-error manipulation on the 2-4-6 task. That would relieve me of the burden of having to prove such instructions had no effect. I had limited resources and could not afford to invest in any study that was not both publishable and directly germane to my interests.

5.2.5. DAX-MED AND THE POSSIBILITY OF ERROR

Despite his aversion to strategy instructions, Tukey had tried Tweney's DAX-MED manipulation and replicated its overall positive effect on task performance. If disconfirmatory instructions did not produce a difference between possible-error and no-error conditions on this task, what about DAX-MED instructions? They would certainly improve baseline performance across conditions, and that might be enough to allow subtle differences to emerge.

In this pilot study, I ran twenty-two subjects on the multidimensional 2-4-6 task, with twelve randomly assigned to no-error and ten to possible-error conditions. The DAX rule was the same as the rule in the two previous studies, i.e., odds and evens must alternate within each triple. The MED rule encompassed any triple that was not a DAX. The possible-error instructions explained that covering a DAX triple meant you thought it really should have been a MED, and vice versa.

Five no-error subjects solved the rule, as opposed to two possible-error subjects. This difference, although in the expected direction, was not significant. No-error DAX-MED subjects also obtained the highest proportion of MED or incorrect triples in any of the studies: .57, as opposed to an also large .49 in the possible-error condition. Furthermore, six of the possible-error subjects actually assigned errors.

Clearly, as in earlier studies by Tukey, Tweney, and Gorman, Stafford, and Gorman, the DAX-MED manipulation produced a high proportion of information disconfirmatory with respect to the DAX rule, although the overall solution rate was still low, about 33 percent. There was also a suggestive, though not significant, difference in solutions between possible-error and no-error conditions. Subjects in the possible-error condition frequently used errors to immunize hypotheses, as illustrated by the following three responses to a question on a brief survey given to all subjects at the close of the task: "Did the possibility of error affect your strategy? If so, how?"

> Yes, because when I saw a triple that didn't fit that didn't really mean my hypothesis was incorrect. I found the possibility of error very frustrating. I assigned errors to the two triples that didn't fit my hypothesis.
>
> Yes, I assigned errors to make my hypothesis true.
>
> I did not assign any errors to the triples because they all seemed to fit my hypothesis.

The first two of these three subjects assigned errors in a way that immunized their hypotheses; the third pursued a confirmatory heuristic with respect to the DAX rule, and therefore did not need to use error to immunize his hypothesis. Possible-error subjects, by and large, failed to follow the methodological falsificationist's advice: only four of ten subjects replicated one triple each; similarly, three of the twelve control subjects replicated at least one triple, one of them replicating two. This observation cannot, of course, be used as evidence that the methodological falsificationist's heuristic is either effective or ineffective; to obtain that sort of evidence, we would actually have to give subjects instructions to employ such a heuristic.

Possible-error subjects proposed an average of 23.4 triples and no-error subjects proposed 34.4—a surprising reversal of the usual result, where possible-error subjects actively pursue replication-plus-extension and therefore propose more triples. This apparent difference was not statistically significant, because of wide variations in the numbers of triples proposed by individual subjects; therefore, the difference itself could have been due to chance, and would itself require replication with more subjects.

Subjects in both conditions followed a holding-constant heuristic on over 60 percent of their triples. In the absence of think-aloud protocols (see Ericsson and Simon 1984) it is impossible to distinguish precisely holding-constant from replication-plus-extension. (Had I the resources, I would have conducted protocols.) But it is clear that in these multidimensional environments, subjects spend a lot of time trying to find out which dimensions really play a role in the rule; this may interfere with their attempts to identify errors. If a subject proposes 1,5,7,B in black, follows it with 1,5,7,B in red and the first is DAX while the second is MED, is the difference in results due to error or to the change in dimension? More replication-plus-extension and holding-constant is required to determine which of these alternatives is correct.

Consider this program of experiments simulating error. When I moved from Eleusis to the 2-4-6 task, I was not holding constant—I varied task and rule, and substituted groups for individuals. But the initial possible-error 2-4-6 experiment was intended as a replication-plus-extension of the Eleusis study. My first multidimensional experiment was both a replication-plus-extension and an example of holding constant, as I changed only one variable: the task. Obviously, these heuristics can be combined to produce a chain of experiments that checks both for error and for which variables are influential.

In fact, my mental representation of a successful research program was still the kind of dramatic difference I had encountered in the Eleusis studies, where every disconfirmatory group had solved the odd-even rule under no-error conditions and only 25 percent had under error conditions. The DAX-MED study showed promise, but solution rates were still extremely low in both conditions. I was searching for a 2-4-6 design that would show strong performance under no-error conditions and weak under possible-error, because that would allow me to contrast my negative or disconfirmatory information with some positive information.

5.3. INCREASING THE COST OF REPLICATION

Adding two dimensions to the 2-4-6 task had clearly increased its complexity, but had it markedly increased the cost of following the methodological falsificationist's advice regarding replication? The use of the term *cost* is consistent with Latour and Woolgar's use of the term *investment;* when considering a replication, a scientist has to decide whether it is a good investment of time, energy, and resources, and obviously the cost of the replication, in terms of these factors, will be a major consideration.

In Eleusis, the cost of replication is high. A card which is right in one location may be wrong in another. So, to replicate a situation where the queen of hearts is correct when played after the ace of spades, subjects have to get to a point in the sequence where the ace of spades is correct, and then play the queen of hearts again. By contrast, even in the multidimensional 2-4-6 task, to check if a triple is an error one merely needs to replicate the individual triple, not the surrounding ones. Thus, the difference between the results obtained on Eleusis and on the 2-4-6 task might be due to the increased cost of replication in the former study.

This problem of cost was particularly obvious to me because it was affecting my own pursuit decisions. Had I a stable of graduate students and a lighter teaching load, I would have pursued the DAX-MED possible-error study. But I sensed that it was not going to be a good investment decision; I would learn less about the possible-error problem from it than from some new design.

One obvious way to increase the cost of replication in these simulations was to limit the number of experiments a subject could perform. Subjects could be told to solve the rule using only ten triples, or even five. This somewhat mirrored my own situation: I had resources to run about two full-fledged experiments a year, and therefore had to choose carefully.

But this seemed to me overly simplistic. While it is true that most scientists have to consider carefully which experiments to conduct next, given constraints on equipment, funding, and professional advancement, they do not make these decisions in a vacuum. New experiments are always proposed in the light of previous results—as well as in a larger social and theoretical context. While I was not ready to simulate aspects of this larger context, I did see a way to simulate review of previous results gathered by others.

5.3.1. A PROCEDURE FOR REVIEWING PREVIOUS RESULTS

After piloting several alternatives on my research assistants—all undergraduate engineering students at Michigan Tech who were helping me as part of an independent study—I settled on the following pattern of previous results.

Triple	Does This Conform to the Rule?
1,2,3	Y
1,2,2	N
5,10,15	Y
10,20,30	N
7,14,21	Y

Each subject in a new experiment was given five minutes to study these triples and write down any guesses he or she had about the rule. Note that a rule based on multiplying the first number in the triple by a constant would be sufficient to explain the pattern of Ys and Ns if 10,20,30 were made an error. Subjects were then shown an additional five triples.

10,33,12	Y
3,23,3	N
8,21,36	Y
12,7,11	N
15,15,15	N

They were given five more minutes to study all ten triples and write down any additional guesses that occurred to them. Note that subjects have been given enough information to discover the rule, which was "odds and evens alternate," and also to discover that other hypotheses, e.g., the one based on multiples, cannot be sustained unless errors are assigned.

Then subjects were allowed to propose five triples of their own, taking as much time as they wanted. They had the original ten triples and their guesses in front of them while they were making their additional choices. They were asked to make a final guess that covered all fifteen triples.

I hoped I had created a design that would increase the cost of replication in an ecologically valid way, though it was clear this sort of review of previous results was far from what a scientist actually does when reviewing the literature. But at least the possibility of error would allow subjects to ignore some previous results. Scientists do something similar when they dismiss certain studies in the literature as methodologically flawed, though again scientists use far more sophisticated criteria for making such judgments than the mere possibility of chance error.

Thirty-two volunteers were drawn from introductory psychology and group communication classes at Michigan Tech and were evenly divided by random assignment between possible-error and no-error conditions. Subjects in the possible-error condition could designate as errors any of the triples, including the ones they were given to study, and cover them with strips of paper, and they could rearrange the error-strips as they chose.

5.3.2. RESULTS

There were no significant differences between possible-error and no-error conditions in terms of solutions or replications. Six of sixteen possible-error subject and eight of sixteen subjects in the no-error condition solved the odd-even rule. Possible-error subjects replicated about 10 percent of the time on their five final triples, and no-error subjects replicated about 5 percent of the time.

To assess the extent to which subjects used replication-plus-extension, I borrowed a design from Tukey (1986), who asked subjects to indicate why they had proposed the triples they did by giving them a list of terms to choose from. Tukey's goal was to make research on the 2-4-6 task relevant to issues other than confirmation or disconfirmation; his terms included *confirm, disconfirm, explore, eliminate, discover, search,* and *assess.* I added *replicate.* The definitions of these terms were brief; my preliminary research suggested that each subject seemed to use these terms in idiosyncratic ways, some labeling each triple with a different term, others using the same term for all five.[11]

But at the very least, these other terms would serve as alternatives to "replicate," the primary focus of my research; subjects would presumably use this term only when there was no other alternative, and that would give me a sense of when they thought they were replicating. Seven possible-error subjects listed a triple as a replication; only four subjects actually proposed the same exact triple at least once. For example, one subject listed 5,7,9 as a replication after 2,4,6; another listed 5,12,33 as a replicate after 1,4,7; another listed 3,6,9 as a replicate, presumably of 5,10,15 in the original set of ten. Others listed repeats of identical triples as replicates. Three no-error subjects listed a triple as a replication; all three repeated the same triple once. So the replication term provides additional evidence that at least three possible-error subjects pursued a replication-plus-extension strategy, though subjects' general confusion over the use of the terms meant that there were probably other instances when this heuristic was employed.

Possible-error subjects did use their yellow strips of paper to designate previous results as errors. Eleven of them made at least one of the first five triples an error; by the end of the second five, twelve had assigned errors; by the end of the experiment only eight possible-error subjects still had errors remaining.

5.3.3. USING COROLLARIES TO IMMUNIZE HYPOTHESES

The possibility of error did permit at least seven subjects to immunize hypotheses in this study, in the sense that they were able to use error to account for potentially disconfirmatory results. However, no-error subjects were able to add corollaries to their hypotheses to accomplish the same goal. To understand the full effect of the possibility of error in the context of reviewing earlier results, it is necessary to study what subjects actually did.

Of the seven possible-error subjects who covered potentially disconfirmatory triples with strips of paper, two guessed that the final rule was "each triple has two odd (or even) numbers and one even (or odd) number." Both assigned errors to the two triples in the original ten that did not fit that pattern; afterward, each proposed triples that appeared to disconfirm this hypothesis and then covered these triples with strips of paper as well.

Two other possible-error subjects hypothesized that the rule was "odd-even-odd" after they had seen all ten of the given triples; both covered the two triples in the original ten that did not conform to their rule, and used a confirmatory heuristic to determine the correctness of their hypothesis.

Two other possible-error subjects had hypotheses involving constant multiples, suggested by the pattern of the first five triples. One said the rule was "take a sequence of integers starting with the odd numbers and multiply it by 1,2,3, respectively." This subject covered the two triples in the original ten that did not conform to his rule and tested his idea with only one disconfirmatory triple, 16,42,72, which was incorrect. Another said "the rows each have a one-digit number in the first column and a two-digit number in the second and third columns"; he had to assign three errors in the first ten triples to make this hypothesis work; although he proposed one disconfirmatory triple in his final five, it happened to have all even numbers and so was incorrect.

Finally, one subject proposed that the rule might be "alternate yeses and nos starting with yes" after the first five. Only the last triple in the first ten failed to follow this pattern, so he covered that with a strip of paper. He proposed two disconfirmatory triples, but when they did not fit his hypothesis, he simply made them errors.

Three other possible-error subjects either came up with no final hypothesis or had only partial hypotheses that they knew were not correct. As one said on his final questionnaire, "the first five conformed to my original theory [which had to do with multiples of 1, 2 and 3], but then I couldn't think of another rule that would fit." So four possible-error subjects were unsuccessful at coming up with any hypothesis that fit their data, even given the opportunity to use errors to immunize hypotheses.

This inability to come up with a rule that fit the data was also characteristic of six no-error subjects, four of whom admitted they had no idea. After the first five triples, a fifth subject's hypothesis was "add the first number to the second to get the third." To explain the fact that 10,20,30 was wrong, she had to add a corollary, or what Holland et al. (1986) call a subordinate exception rule: "if the result was even, it doesn't fit the rule." The next set of triples disconfirmed even this modified hypothesis. So she added another corollary: her hypothesis worked only on an alternating basis, i.e., for every other triple. Even so, it does not explain all the data she generated.

This subject illustrates why control subjects performed as poorly as did possible-error subjects; the former used corollaries to immunize their hypotheses. Consider another subject, who proposed that the rule was "last number divided by the first is greater than the last number divided by the second" but added the corollary that this relationship reverses when the direction of the triple reverses. This subject followed Platt's strong-inference heuristic, eliminating alternatives like "even-odd-even" and "multiplying by a constant" before arriving at this guess, which still did not fit at least one triple in the original ten—10,20,30—and also failed to fit several triples that the subject generated.

The two subjects above used corollaries unsuccessfully: final guesses still did not fit their data. In the absence of any alternate hypotheses, they could not let go of their "hard core" ideas about the rule. Three other control subjects successfully immunized final hypotheses with corollaries. One of these subjects immunized "multiply by a constant" with a corollary: "if a digit is repeated in the second column the third number is smaller, but larger than the first number." This was his way of accounting for the triple 10,33,12, which he could have labeled an error in the possible-error condition. One of his final five triples—9,88,50—attempted to confirm this corollary but was incorrect. So, he added a corollary to his corollary: the second number with the repeated digit had to be odd. Again, he might have made this triple an error in the possible-error condition.

The two other no-error subjects had final guesses that were several paragraphs long. For example, one created separate rules to cover what happened if a triple were even and greater than or equal to ten, and if it were even and less than ten; in the latter case, his rule read, "the second number is three times the integer preceding the first number and the third number is four times the integer immediately following the first number." His final guess was a set of such corollaries and resembles Lakatos's idea of a degenerating research program: increasingly desperate attempts to deal with data that contradict any existing hypothesis.

Therefore, the possibility of error did give several subjects a way to immunize hypotheses on this task, but limiting the number of experiments had an unanticipated effect—in Lakatosian terms, it made it easier for no-error subjects to immunize their hypotheses by coming up with "protective belts" of corollaries designed to fit every potential disconfirmation. In some cases these additional corollaries began to resemble a degenerating research program.

Clearly, limiting the number of triples—and therefore replications—that subjects can propose is not sufficient to reproduce the performance differences noted when the possibility of error was added to Eleusis, even though the cost of replication made it harder to follow the methodological falsificationist's advice: less than half of the possible-error subjects replicated or used replication-plus-extension. Instead, providing subjects with results to study and limiting the triples they could propose encouraged hypothesis-perseveration in both conditions: possible-error subjects used error to immunize hypotheses, and control subjects used corollaries.

5.4. WHEN CURRENT RESULTS DEPEND ON PREVIOUS ONES

My mental representation for a "successful" result in this series of studies was still the kind of dramatic difference I had seen when the possibility of error was added to Eleusis. The latest experiment had revealed a phenomenon of substantial interest—that reviewing a small body of data and conducting a limited number of experiments encouraged some subjects to embark on a degenerating research program. Increasing the cost had also reduced the number of replications. But I was still concerned that I needed a positive result, both to ensure publication and to pinpoint why the Eleusis possible-error study worked.

I continued to consider the difference in the cost of replication on Eleusis and the 2-4-6 task. Imposing a limit on the number of 2-4-6 replications was not precisely identical to the problem encountered on Eleusis. The odd-even rule in Eleusis began with an odd card and continued even-odd-even throughout all the cards played. The rule in the latest experiment merely specified that odd and even numbers had to alternate within a triple, but not across the whole set of triples. Replication is more difficult in the former rule because one cannot just replicate a single experiment or trial, one must replicate the previous experiments as well.

Therefore, the possibility of error might have its greatest effect in situations where results of the current experiment or trial depend on results of the previous one. A scientific analogy might be the "great Devonian controversy," as described by Rudwick (1985). Participants in this controversy were attempting to determine the sequence in which geological strata had been laid down, over time. During the course of the controversy, it became apparent that the most reliable way to identify a given stratum was to look at where it fit in with other strata above and below it, using fossil records as well as geological characteristics.

In other words, whether a given observation supported or disconfirmed a geological hypothesis depended on what observations were seen as preceding and following it. Competing geologists De la Beche and Murchison argued about both the sequence and the nature of strata in Devon; what was an anomaly for one man's theory often provided support for the others. The consensus that emerged was closer to Murchison's original views than to De la Beche's, but Murchison's changing hypotheses were affected by anomalies uncovered by De la Beche (see Rudwick 1985: 412–414 for a summary). Similarly, in Eleusis, the only way to predict whether a card will be correct or incorrect is to know the sequence of cards in which it appears.

Like groups in Eleusis, participants in the Devonian controversy spent a great deal of time debating whether and where errors occurred. One of the standard defenses against an apparent anomaly is to label it an error. Murchison labeled De la Beche's original sequence of strata an error, focusing particularly on one stratum, and made a field trip to Devon in which he claimed to find a different sequence. This led to other field trips by other geologists, often going over this same Devon ground to see if errors could be found in earlier sequences. Some of these field trips resembled replications, in that the goal was to see if a particular ordering could be seen by other observers; other trips resembled a kind of replication-plus-extension, an attempt to see, for example, if a sequence in Devon could be reproduced in the Ruhr Valley in Germany (see Rudwick 1985: 304–306).

Another example of observations that depend on sequence comes from biology. Keller (1983) describes how Barbara McClintock discovered transposition of genetic material in chromosomes. The effects of this transposition were first observed as occasional colored spots on corn kernels that should have been white. McClintock's search for the genetic mechanisms underlying this minor aberration—which could have been dismissed as noise or error—led her to a complex rule that specified how a piece of chromosome can break off and insert itself in a different place. This transposition would have to occur early in development to have the observed effect, which explained why most ears do not show these colored spots. "The variegation patterns that had led her down this long and arduous road were not, as some might think, expressions of a breakdown of normal processes. Rather, they could be seen as 'merely an example of the usual process of differentiation that takes place at an abnormal time in development'" (Keller 1983:135). In other words, McClintock came up with a hypothesis which specified that an apparent aberration, or "error," in the appearance of kernels of corn was really evidence for an underlying regularity of major importance.

When sequence is a feature of the rule, the methodological falsificationist's heuristic is particularly hard to follow, because replication of an entire sequence of experiments or observations is necessary to establish whether a disconfirmation is an error. In the Devonian example, replication was done by competing research teams who checked each others' sequences from different perspectives until a consensus emerged. In McClintock's case, she did the replications herself—painstakingly, over many years. But she also had an intimate knowledge of the genetic material she was working with, to the point where she knew all the potential sources of errors and could eliminate them. In other words, she may have been able to reduce the possibility of error to zero through her methodological expertise. Future experimental simulations should give subjects a way to refine measurements or procedures before replicating an experiment. As it stands, Eleusis and 2-4-6 subjects can only use replication to check for error.

There are other problems with drawing an analogy between geological strata, genetic sequences, and Eleusis. The latter task simulates experimental reasoning; subjects must invent their own trials, not observe preexisting sequences of the sort encountered in the former cases. When subjects are given trials to review, as in the version of the 2-4-6 task used in the last experiment reported above, the analogy to the Devonian and McClintock situations is stronger: now subjects must infer a pattern from observations, then generate new evidence to test it. If results on the ten original triples depended on where they fit in a sequence, perhaps the possibility of error would have a greater effect on performance: a result that was out of sequence could be dismissed as an error, and it would be harder to employ the methodological falsificationist's heuristic to check for errors.

5.4.1. Adding Sequence to the Rule

In my next experiment, I decided to make a small, but significant, change in the odd-even rule: instead of numbers alternating odd-even just within the triple, the alternating pattern would have to continue across triples as well. This would make the 2-4-6 rule more like the one I had used on Eleusis. Therefore, the ten starting triples were altered:

Triple	Does This Conform to the Rule?
1,2,3	Y
4,5,6	Y
4,5,6	N
5,10,15	Y
10,20,30	N

As in the previous study, subjects were given five minutes to write down any hypotheses they had about the rule at this point. Half of the twenty-two subjects, randomly selected, were run in a possible-error condition in which they were permitted to designate any of the triples as errors and cover them with strips of paper. Then, subjects were given a second set of five triples:

10,33,12	Y
13,20,5	Y
14,9,14	Y
12,35,14	N
15,15,6	N

Again, they were given five minutes to write down hypotheses covering all ten triples. Note that a simple "odds and evens alternate within the triple" rule identical to the one used in the previous experiment could be maintained by covering triples three (4,5,6) and nine (12,35,14) with strips of paper. The triple 4,5,6 in the first five was correct after 1,2,3 because its first number (4) was even, following an odd number (3) at the end of the last correct triple. But this same triple was incorrect when repeated because it put two even numbers (6 and 4) back to back, breaking the pattern. I expected that possible-error subjects would make the second 4,5,6 an error, whereas control, no-error subjects would have to explain why replication produced inconsistent results.

Finally, as in the previous study, subjects were given unlimited time to propose five additional triples, then were asked to write a final guess that covered all fifteen.

5.4.2. RESULTS

Finally, I achieved a result that matched my expectations. The possible-error manipulation had a dramatic effect on performance: Three of twenty-two possible-error subjects solved the rule, as opposed to half (eleven) of the no-error subjects. There was no corresponding difference in the amount of replications: on the average, both possible-error and no-error subjects proposed one triple in their final five that was an exact duplicate of an earlier triple. Apparently, in this situation, subjects in both conditions followed the methodological falsificationist's heuristic with equal facility.

What about replication-plus-extension? Subjects were again asked to write one of the seven terms suggested by Tukey (1986) or an eighth term (replication) after each triple. Again, use of Tukey's terms fell into no clear pattern, but I felt the use of the replication term might give indirect evidence of replication-plus-extension. Unfortunately, subjects in this study used the term more literally: in both conditions they wrote the word *replication* after a triple just about exactly as often as they proposed the same triple twice: thirteen possible-error subjects used the term and fourteen actually replicated; ten no-error subjects used the term and thirteen actually replicated. The term was used often after repetitions of triples in the original ten, e.g., 4,5,6 and 10,33,12, but there were times when a triple was repeated and the term was not used. For example, one possible-error subject proposed 10,33,12 twice, labeling the first attempt "replication" and the second one "discovery." Another possible-error subject labeled all five of his final triples "explore," including a repetition of 4,5,6. One possible-error subject who actually replicated on all but one of his final triples labeled only one of these a replication. Incidentally, this subject solved the rule, suggesting the power of the replication heuristic.

Occasionally, the term *replication* did spot an instance of replication-plus-extension—e.g., one subject who proposed 9,10,19 to replicate 7,8,15—but in general, Tukey's terms were too ambiguous for subjects to use them properly. Future studies should abandon their use and find better ways of accessing what heuristics subjects were employing (see note 11).

This 20 percent replication rate was too low to eliminate the possibility-of-error factor: all but four possible-error subjects still had one or more triples covered with yellow strips by the end of the experiment. But did possible-error subjects engage in what Kern (1982) had called "hypothesis perseveration"? Five certainly did: beginning with the hypothesis that the rule was simply "numbers in the triple must alternate odd-even-odd or even-odd-even"; these subjects made triple numbers three (the second occurrence of 4,5,6) and nine (12,35,14) into errors. Had they not made these triples errors, their hypothesis about the rule would not have been consistent with the evidence. All five of these subjects followed a confirmatory heuristic on their final triples; only one inadvertently obtained a disconfirmatory result, but dismissed it as an error.

Four other possible-error subjects also engaged in hypothesis-perseveration; in addition to assigning errors, they added corollaries to immunize odd-even hypotheses. For example, one subject who made triple nine an error concluded that the rule was "odds and evens alternate within the triple," with some additional corollaries—a certain triple pattern could only appear once, and the total of the numbers in a triple added together could not exceed 50. This allowed him to deal with the fact that when he replicated 10,33,12 it was wrong, though it had been right in the original ten.

Four possible-error subjects had final hypotheses that explained only a portion of the evidence. For example, one said the final rule did not permit repetition of triples or negative numbers, but could not explain why some triples without these characteristics were incorrect.

The possibility-of-error also made it impossible for six subjects to come up with any hypothesis. In response to the question "What strategy did you follow on this task?" one of these subjects, who worked for two hours, said:

> At first I thought there was a consistent rule to follow. I think I looked too heavily at the random error possibility. With that option always looming, I never had total confidence in trying to develop any rule. There were always exceptions and I could never be sure. I formed a new rule every time one of my triplets was rejected.

Contrast this response to that of the lone subject in the no-error condition who had no final guess. He wrote the following in the space where he was to indicate the rule for all fifteen triples:

> I can't think that the rule had anything to do with the triple itself because '4,5,6' was listed three times [the third time by the subject himself] and first it received Y, then N, then N. I feel the rule deals with either the letters of the numbers *or* the rule involves *previous* triples.

In other words, whereas the possible-error subject discussed above was overwhelmed by the number of possible rules, this no-error subject developed an appropriate mental representation of the type of rule he was looking for, i.e., one involving sequence, but could not come up with any hypotheses based on that representation that fit his data.

In general, this was the problem that faced almost all the no-error subjects that did not solve the rule: none of their final hypotheses fit their data. The lone exception was a no-error subject who said the rule was "the first and last numbers must both be even or odd" and added the corollary, "the triple cannot be a multiple of the preceding triple." This hypothesis actually fits the data he generated, though of course he failed to test adequately the multiples corollary.

Two no-error subjects had partial guesses. One's final hypothesis read, "Any triple cannot replicate another triple while not exhibiting any recognizable relationship to any other." In effect, this "not exhibiting any recognizable relationship" is his admission that he could find no other pattern.

Eight other no-error subjects did come up with final hypotheses, but had to ignore some evidence to make these hypotheses work. One subject whose final guess was "the odd triples cannot have a middle number less than five and all the numbers in the triple must add up to less than 60" admitted on his questionnaire that "for some reason my mind blocked out triple 14 [12,35,14] for awhile so my strategy would have been different if I realized my proposed rules didn't explain the failure of triple 14." Another proposed the rule that "the difference between the numbers in the triple is a prime integer" and added the corollary "they can not be repeated," but this hypothesis is contradicted by the fact that the triple 13,20,5 in the original ten is correct. Also, his use of "they cannot be repeated" is ambiguous; the "they" appears to refer both to repeating numbers within the triple and to repeating a triple.

5.4.3. Conclusions

The methodological falsificationist's statement that additional replications are sufficient to eliminate error assumes an ideal situation where unlimited experiments are possible. The experiment above suggests that, in situations where only a limited number of experiments can be performed and results of the current experiment or observation are dependent on where it fits in a sequence of experiments or observations, even a replication rate as high as 20 percent is not sufficient to eliminate the possibility of error. The possibility of error either encouraged subjects to immunize hypotheses by dismissing disconfirmations as errors or so overwhelmed them that they were unable to come up with a final guess. In contrast, no-error subjects either guessed the necessary and sufficient rule or came up with hypotheses that failed to account for all the evidence.

Similarly, in the Eleusis experiment reported in chapter 4, the possibility of error often permitted groups to immunize their hypotheses against disconfirmation. But those groups specifically given disconfirmatory instructions realized that there was no actual error; however, they spent so much time replicating potential errors that they failed to obtain information that would have disconfirmed rules like "adjacent cards must be separated by ones" that were subsets of one of the actual rules, "odd and even cards must alternate." Possible-error subjects in the study reported here were constrained by the original ten triples and by the limit on the number of additional trials they could perform; therefore, they had to deal with evidence inconsistent with any rule besides "numbers must alternate odd-even within and across triples."

About half of the no-error groups working on Eleusis solved the odd-even rule, across strategy conditions. Similarly, half of the no-error subjects in the present study solved a similar rule. However, Eleusis groups who failed to find the experimenter's rule were able to come up with hypotheses consistent with their data, again because they were not constrained by the ten original triples that would have forced them to consider evidence not consistent with their hypotheses.

No-error subjects given previous results to study were able to immunize their hypotheses through the use of corollaries when sequence was not a feature of the rule, but once sequence was added, even those no-error subjects who used corollaries like "the same triple cannot be played more than once" still had to ignore results inconsistent with their hypotheses. Why didn't corollaries work on this version of the task? Apparently the sequence feature simply made it too difficult to fit any combination of hypotheses and corollaries except the correct one. In particular, even many of the unsuccessful no-error subjects developed an appropriate mental representation of the type of rule, recognizing that sequence played a role, but could not translate this representation into a successful hypothesis.

In contrast, most possible-error subjects did not recognize the sequence feature. Eleusis subjects knew that the rule involved sequence, so the possibility of error in that study did not interfere with rule representation; instead, it seemed to affect use of a disconfirmatory heuristic.

These results suggest that in scientific situations where sequence rules are expected, the possibility of error will encourage hypothesis-perseveration. This kind of situation may be encountered more often in biology and geology than in physics, as the Devonian and McClintock examples suggest. Note that disconfirmation in the Devonian case was achieved by competing research teams, each seeking to discredit the other.

But there are experiments in physics where sequence is important, too: consider a chain of elementary particle reactions, where the interpretation of the final result depends on the whole sequence. Similarly, results in chemistry often depend on long chains of reactions; an error at any point in the sequence will alter the final result.

These results would also lead us to predict that in a novel situation where scientists did not expect a sequence rule, the possibility of error would make it easier to dismiss that sort of pattern. An example might be the initial discovery of sea-floor spreading (Giere 1988). A graduate student named Vine took magnetic data from a section of the mid-Atlantic ridge and, because he possessed the appropriate "cognitive resources," to use Giere's term, he determined that there were "blocks of alternately normal and reversely magnetized material" moving away from the center of the ridge and parallel to its crest. These cognitive resources included procedural knowledge, e.g., the ability to analyze this data on a digital computer, but also a mental representation based on familiarity with theories of sea-floor spreading.

This new "Vine-Mathews hypothesis" did not win immediate acceptance. A competing laboratory, which took the view that the oceans were stable, dismissed Vine's results as a kind of error, in that most other ridge profiles were not similar. Now numerous other factors besides the initial mental representation play a critical role in this controversy, but whether one regarded the ridges as spreading sequentially or as static features clearly determined how one initially viewed the data.

Of course, sequence rules and strips of paper on the 2-4-6 task are a long way from the situations faced by geologists and biologists. But these artificial tasks do allow us to make externally valid generalizations about the circumstances under which the mere possibility of error can interfere with problem solving. I felt I had finally solved the riddle of the apparently inconsistent results, not only by confirming that adding sequence to the 2-4-6 task increased the possible-error effect, but also by disconfirming the role of other factors like multidimensionality.

Now it was time to publish. I assembled the four main experiments described in this chapter into a manuscript, omitting the DAX-MED and disconfirmatory-instruction pilot studies, and shipped them off to a major American journal, in lieu of the *Quarterly Journal of Experimental Psychology* and the *British Journal,* my publication mainstays. I felt it was time to try to reach an audience in my own country.

While I waited the three to six months it took for a response, I followed my usual heuristic and embarked on the next study in the research program. My first experimental paper had been accepted, in part, because I had followed this heuristic—I was able to add preliminary results from the 2-4-6 task to demonstrate that my Eleusis results were not simply due to chance error or peculiarities of the task.

But what was the next step? It seemed obvious to me. Armed with an understanding of the effect of the mere possibility of error, I should turn to the effect of actual error, and I should do it by returning to my baseline task: Wason's original 2-4-6. This experiment and the relevant theoretical literature will be described in the next chapter. We will return to the story of my attempts to publish this research program in the chapter after.

At this point in my research program, I felt I had found the conditions under which the possibility of error had its greatest effect: when the results of current experiments depended on where they fit in a sequence of previous results. This type of rule made replication and replication-plus-extension particularly difficult; to determine whether and where error had occurred, one had to replicate an entire sequence of events, not just a single trial. Therefore, if one thought the target rule might contain a sequencing component, one would have to follow a "replicate-sequence" heuristic.

Application of this heuristic depends on how the task or problem is represented. What constitutes a "replication sequence"? That will vary from problem to problem and, possibly, from scientist to scientist. Once again we have an externally valid finding whose ecological implications need to be checked carefully. The Devonian and genetic transposition cases are merely suggestive (see 5.3). What is needed is a more detailed case study of how scientists check for error in a situation where they think the principle or pattern involves a sequenced interdependence of results. More experiments with different sorts of tasks involving sequence rules would also shed light on this matter.

But, for my next experiment I wanted to move from possible to actual error. Would replication suffice to detect actual all-or-none or system-failure errors in situations where it eliminated the possibility of error? Would subjects be more or less likely to replicate errors that appeared to falsify their hypotheses?

SIX

Simulating Actual Error

6.1. ELECTRONS AND ERROR

In 1913, the physicist R. A. Millikan published his measurement of the charge on the electron, using the famous oil-drop procedure which he had developed. The paper contains data for 58 oil drops; he emphasizes that: "It is to be remarked, too, that this is not a selected group of drops, but represents all of the drops experimented on during 60 consecutive days." Four years later, in his book *The Electron,* he repeated the same theme: "These [58] drops represent all those studied for 60 consecutive days, no single one being omitted" (Holton 1986:9). Here, then, is the perfect, error-free series of experiments, leading inexorably to the conclusion that the electron has a unitary charge.

In reality, a total of 140 drops were observed, 82 more than Millikan reported; these 82 were labeled errors by Millikan and were discarded. For example, right after one of the 58 successful runs, Millikan conducted an experiment which gave values for the charge of the electron that were too far apart. In his notebook Millikan remarked, "Error high—will not use" (Holton 1986:11).

> The "high" error was first of all a judgement stemming from Millikan's presupposition that the smallest charge in nature could not be a fraction of the charge of the electron. . . . Millikan's decisions seem to us now eminently sensible; but the chief point of the story is that, in 1912, Millikan's assumption of the unitary nature of the electric charge was by no means the only one that could be made. On the contrary, a chief reason for his work at the time was to perfect his method and support his claim against the constant onslaught of Felix Ehrenhaft and his associates who, for a couple of years, had been publishing experiments in support of their own, precisely opposite presupposition, namely, in favor of the existence of subelectrons. (Holton 1986:11–12)

In other words, Millikan's mental representation told him which results were erroneous and which should be taken seriously. In Lakatosian terms, Millikan's presuppositions about the unitary charge on the electron may have served as a kind of "hard-core" assumption which could not be disconfirmed; when results did not accord with expectations, they had to be explained using corollary hypotheses or subordinate rules (see Holland et al. 1986). One obvious corollary explanation for an apparent disconfirmation is to invoke a procedural error, especially in a situation such as this where the procedures were so delicate and required such skill.

Note that Millikan's response to errant oil drops is reminiscent of Einstein's response to Kaufmann's apparent disconfirmation of Special Relativity. In both the Einstein and Millikan examples, the possibility of error played an important role in saving hypotheses from apparent disconfirmations, and Einstein and Millikan turned out to be right—they had identified actual errors. Had Kaufmann's experiment been replicated by other scientists and Millikan's not, the eventual outcomes might have been different.

Consider the role of replication in the case of the N-ray (Klotz 1980). The distinguished French physicist P. R. Blondlot announced in the wake of the discovery of X-rays that he also had discovered a new type of radiation, the "N-ray." Researchers in France initially confirmed Blondlot's results, but researchers in the United States, Britain, and Germany had trouble replicating Blondlot's discovery. Blondlot attributed their failure to replicate to procedural errors. In fact, he dismissed one of his own disconfirmatory results as an error. When pictures of a spark with or without N-rays were taken, Blondlot predicted that the picture with N-rays would come out brighter. It did not, and Blondlot attributed this to a "poorly regulated spark."

Eventually, the American physicist R. W. Wood went to Blondlot's laboratory and surreptitiously removed the aluminum prism that was supposed to be producing the rays. On the other side of a screen, Blondlot still saw evidence for the rays—even after the element that produced them was no longer present (Latour 1987:75). In this manner Wood successfully undermined Blondlot's claim, though Blondlot himself maintained his faith in the new ray until his death—a classic example of what Kern (1982) calls "hypothesis perseveration."

Obviously, Wood's demonstration played a critical role in disconfirming the N-ray hypothesis. But the methodological falsificationist would point out that the repeated failures to replicate Blondlot's experiments created an atmosphere of doubt, at least on the part of the international community. Is replication sufficient to distinguish cases of legitimate hypothesis-perseveration, e.g., Einstein and Millikan, from cases like Blondlot's where the hypothesis ought to be falsified?

Adding actual error to the 2-4-6 task might permit us to approach an externally valid—but not necessarily ecologically valid—answer to this question. The triples in the 2-4-6 task can be viewed as corresponding roughly to the trials in Millikan's oil-drop experiment: each triple, like each oil drop, can be seen as either supporting or contradicting a particular hypothesis. Millikan knew he faced the possibility of error, and was able, in hindsight, to determine where system-failure errors occurred, system failure in the sense that error made a measurement which otherwise would have supported Millikan's hypothesis appear to disconfirm it. Therefore, if subjects had to discriminate between possible and actual error on the 2-4-6 task, they would be in a situation somewhat analogous to Millikan's.

Of course, there are major differences between oil drops and triples: Millikan's oil drops could only be produced by someone possessing considerable technical skill; triples can be produced by anyone who has a rudimentary familiarity with numbers. Also, whereas subjects in the 2-4-6 task experience only all-or-none error, Millikan's oil drops could also be "thrown off" by procedural or measurement errors which Millikan could then correct. In other words, for Millikan replication involved very precise duplication of certain techniques, and he and his technician's intimate familiarity with the apparatus no doubt contributed to his judgments about error (see Holton 1978 for a detailed description of Millikan's measurement techniques).

College students also do not have the deep theoretical commitments of working scientists. Ehrenhaft went through one of Millikan's early papers and "subjected the data to a devastating attack, turning them against Millikan" (Holton 1978:58). He recalculated the separate charge for each of Millikan's drops, contra Millikan's process of lumping several runs to obtain an average; instead of the data pointing toward a single electron, Ehrenhaft's recalculations made it appear that there could be a wide range of charges, consistent with his theory of subelectrons. "It appeared . . . that the same observational record would be used to demonstrate the plausibility of two diametrically opposite theories, held with great conviction by two well-equipped proponents and their respective collaborators" (Holton 1978:58). Millikan almost immediately published a replication of his earlier results with an improved measurement technique; eventually, Millikan's position was accepted by the scientific community, though Ehrenhaft never abandoned his hard-core belief in subelectrons.[12]

Collins goes so far as to argue that "the value given to a single experiment, positive or negative, seems to depend upon the prior propensity of scientists to believe in the phenomenon in question" (1985:45). In particular, Collins argues that what constitutes a replication is a matter for social negotiation within the relevant communities of scientists. While we can be precise about what constitutes an exact replication on the 2-4-6 task, it is much harder to be as precise in the case of an actual scientific experiment.

One could go on listing ways in which the 2-4-6 task and Millikan's oil drops are and are not analogous. The point is, the 2-4-6 task is not a precise replica of the Millikan situation and it uses college students instead of scientists; therefore, although it possesses some ecological validity, the amount (if one could measure amounts of validity on some imaginary scale) is low. But one can draw externally valid conclusions about whether, on an abstract task, science and engineering students will be able to distinguish between possible and actual error, perhaps by following the methodological falsificationist's heuristic, or whether most will use error to engage in hypothesis-perseveration. The strength and weakness of this kind of simulation is that it eliminates both the expert procedural knowledge and the mental representations of a Millikan; however, this means that we can study how the students form their own representations and develop their own heuristics. Whereas it is impossible to pinpoint the exact effect of all-or-none error in the Millikan example, because error interacts with so many other factors, in the laboratory one can isolate a specific type of error and study its effects. Therefore, I decided to add actual error to the 2-4-6 task and compare its effects to those of possible error.

6.2. PREVIOUS EXPERIMENTAL SIMULATIONS OF ERROR

Although the research reported in chapters 4 and 5 is the only work on the possibility of error, a number of experimental paradigms have been used to explore actual error in results. Particularly relevant to my work was a series of studies done by members of my "invisible college" at Bowling Green, although—as often happens within a small group of researchers pursuing similar problems with similar methods—my first actual-error study was done concurrently with the program of research described below.

Following up on Kern's work, Doherty and Tweney (1988), in a report prepared for the Army Research Institute, described a series of experiments devoted to the two kinds of error originally described by Kern (1982): measurement error and system-failure error. Measurement error corresponds to the normal deviations from the mean that produce a bell-shaped curve. System-failure error corresponds to the type of error whose possibility was introduced in chapters 4 and 5. Doherty and Tweney focus on its ramifications for technological systems; they describe it as an "output that is fundamentally unrelated to the process that the system is supposed to be representing" (p. II), i.e., a kind of malfunction which renders some bits of information useless. Just as a scientist has to ignore certain erroneous results, as Millikan did when oil drops did not correspond to his expectations, so the user of a technological system has to discard certain pieces of information that may be produced by system failures. Doherty and Tweney also investigated whether the error was produced at the input or the feedback phase; the studies in chapters 4 and 5 focused on the possibility of feedback errors only.

One of the paradigms used by Doherty and Tweney is referred to as "multiple cue probability learning" or MCPL. The subject must learn to predict a criterion from a small number of cues. The multidimensional 2-4-6 task could be altered to serve as an MCPL task. Each of the dimensions would have to be related to the rule in a probabilistic fashion, and each might have a different weight. For example, one might specify that the rule was "numbers must alternate odd and even or colors must alternate red and black" and add a probabilistic element: .8 of the odd-even and .6 of the alternating color triples would be correct. The letter, in this case, would remain a distractor dimension. Note that there is no provision for labeling triples erroneous in this design, even though the probabilistic nature of the dimensions has a similar effect, i.e., 20 percent of the time the odd-even rule won't work.

Doherty and Tweney began with a simpler MCPL task. Subjects were given two vertical bars (A and B) on a computer screen, and were asked to predict the height of a third bar (C). The multiple cues, in this case, were the heights of the two predictor bars; the correct solution was to use the average of A and B to estimate C. In the measurement (ME) error condition, a normally distributed random component based on the multiple correlations between A, B, and C was added to the sum of A and B. In the system-failure (SF) condition, a value for C was selected at random from the distribution of the averages of A and B. In other words, when A and B each had heights of one, in the ME condition the height of C would be close to one but in the SF condition it might be a value highly divergent from one. SF subjects were able to ignore the error trials and make accurate predictions of the criterion. ME subjects performed slightly, but not significantly, worse than SF subjects. This result appeared to contradict Kern's (1982) earlier findings.

In a follow-up study, the MCPL task involved using water and fertilizer levels to predict plant growth. SF error was set at 30 percent; ME error was again a small, normally distributed value added to the actual growth. In this study, SF error was far more disruptive than ME error, which replicates Kern's earlier result. This MCPL task was more complex than the one used in the previous study, which suggests that the effect of SF error depends, in part, on task complexity.

6.2.1. Error in an Artificial Universe

Klayman (1988) conducted an MCPL study using a scientific reasoning task, based on the artificial universe developed by Mynatt, Doherty, and Tweney (1978). Klayman's subjects had to combine such variables as shape and size and shading of target to predict the distance a particle would travel. In one experiment, he added about 20 percent random SF error to this universe. Subjects, working over several days, were still able to discover the relevant cues in this condition, although "they were less likely to express generalized rules of system behavior" (Klayman 1988:328). In other words, they had trouble coming up with theories that explained how the universe worked, but they did at least know what variables were relevant. Recall that this version of the artificial universe was so complicated that no subject solved it even when there was no error (Mynatt, Doherty and Tweney 1978). Similarly, most subjects in the multidimensional 2-4-6 task were able to determine that only the number dimension was relevant to the rule, even though few subjects were able to solve the rule.

Klayman's major concern was whether subjects would even be able to discriminate the relevant dimensions on an MCPL task. Faust (1984) describes research on clinical judgment which demonstrates that both experts and novices working in ecologically valid situations are not so discriminating; they maintain faith in the validity of cues that are not predictive. "For example, bats and morbid thoughts are often linked together. Therefore, if a Rorschach response contains such verbal content as 'It looks like a bat,' individuals naturally assume that the respondent has morbid preoccupations" (Faust 1984:68). One study by Chapman and Chapman (1969) involved three levels of random error—33 percent, 17 percent, and 0 percent—in the relationship between Rorschach cards and personality traits. Subjects were able to perceive the correct relationship only in the 0 percent condition, because they focused on cues for illusory relationships, as in the bat example above. Faust argued that in these sorts of judgment situations, where prior beliefs play an important role, subjects—and even professionals—tend to follow a confirmatory heuristic. When prior beliefs are not as important, subjects show more willingness to use disconfirmatory information. In probabilistic environments, as in situations where there is error in data or information, beliefs and expectations play an even more dominant role than they do in situations where relationships are 100 percent reliable. Consider the Millikan and Blondlot examples: their prior beliefs helped them decide what constituted an error and what constituted a valid result.

But what is the effect of SF error in situations where multiple cues are not involved? Mathews, Buss, Chinn, and Stanley (1988) asked subjects to discover a concept related to a series of Chinese ideographs. Four levels of random error in feedback were introduced: 0, 20 percent, 40 percent, and 60 percent. The subjects were not warned about the possibility of error. Mathews et al. found that as the amount of error increased, subjects showed an increasing tendency to engage in hypothesis-perseveration, i.e., high proportions of feedback error encouraged subjects to repeat the same hypotheses rather than generate new ones. Mathews's error corresponds to the SF error used in the Kern and Doherty and Tweney studies; therefore, Mathews's result supports Kern's original observation that SF error causes hypothesis-perseveration.

Doherty and Tweney conducted two follow-up studies directed at distinguishing between measurement error and system-failure error using Kern's artificial universe, which is not an MCPL task. They sought to improve Kern's original study by clarifying the difference between type of error and locus of error. To review, Kern's subjects dropped probes on the surface of a planet to determine the boundary between "live" and "die" zones for "tribbles." Kern's measurement error had to do with an uncertainty in where the probe was located; her system-failure error lay in the feedback from the probe, and caused hypothesis-perseveration.

In the first experiment, Doherty and Tweney placed both measurement (ME) and system-failure (SF) error in the location of the probe, and in the second, they put the locus of both types of error in the feedback from the probe. In this first study, they crossed three levels of ME error, none, low and high, with three of SF error, 0 percent, 25 percent, and 50 percent; in the second study, they used only low and high levels of ME error and 0 and 30 percent levels of SF error.

Both kinds of error significantly affected performance in the first study, but only ME error did so in the second. Overall, Doherty and Tweney concluded that, "Whereas SF error caused a small proportion of subjects to do very poorly, increasing levels of ME error appeared to cause a large number of subjects to perform slightly less well" (1988:186).

Kern's and Mathews's observations that SF error caused hypothesis-perseveration was not replicated in these studies; most subjects in the SF error conditions revised their hypotheses as readily as did subjects in other conditions. Nor were there any differences in replications. Doherty and Tweney concluded that "Kern's study found large and consistent changes in the strategies used by subjects in the presence of SF error. We found less strong effects. It is not clear why this difference exists, but it is plausible to suggest that the need to unconfound locus of error and type of error led to a more complex task environment for our subjects" (1988:193–194). In other words, in these experiments, Doherty and Tweney found that SF error did lead to performance problems on the part of a few subjects but did not lead subjects to adopt a hypothesis-perseveration strategy. As in the possible-error studies, characteristics of the task become very important in explaining performance.

6.3. SYSTEM-FAILURE ERROR AND THE 2-4-6 TASK

To explore the interaction between task and error, Doherty and Tweney, in collaboration with Bonnie Walker (Walker, Doherty, and Tweney 1987), conducted two experiments using the 2-4-6 task, which they computerized. In the first, they compared three system-failure error conditions: (1) no error, (2) informed error, and (3) uninformed error. In both the informed and the uninformed conditions, 20 percent random error was programmed into the computer; in the informed condition, subjects were told about it and in the uninformed they were not. Two additional conditions were crossed with the three types of error: subjects were given either DAX or non-DAX feedback on each triple, or DAX-MED feedback (see section 3.4). Wason's traditional "ascending numbers" rule was used; a triple like 3,7,1 would be labeled non-DAX by the computer in the former feedback condition, and MED by the computer in the latter feedback condition. A subject in one of the three error conditions was also assigned to either the one-rule (DAX/non-DAX) or the two-rule (DAX-MED) condition.

Eleven of sixty subjects across both error conditions eventually solved the rule, as opposed to twenty-seven of thirty no-error subjects; this is a highly significant difference, and was paralleled by a comparable difference in the number of first-announcement solvers. (Recall that it is the practice of the Bowling Green researchers to give their subjects feedback on each hypothesis announcement; see section 2.4.2.) Subjects in the two error conditions also proposed almost twice as many triples, on the average, as did no-error subjects, and replicated about twice as often, though much of this difference came from subjects in the one-rule condition; subjects in the two-rule condition did not differ across error conditions in terms of proportion of replications.

The major difference between the one- and two-rule feedback conditions lay in subjects' predictions: one-rule subjects predicted about 60 percent of their triples would be DAX, whereas two-rule subjects predicted about 38 percent would be DAX. This suggests that the two conditions represented the rules differently, though Doherty and Tweney do not explore this ramification. Both feedback conditions obtained roughly similar amounts of disconfirmatory information.

According to Tweney and Doherty, "error-free and error-prone data [should] elicit different hypothesis testing strategies" (1983:153). Specifically, these authors expected subjects operating under informed error conditions to be more likely to follow a confirmatory, as opposed to a disconfirmatory, heuristic. But contrary to expectations, error and no-error conditions did not differ in terms of proportion of triples predicted DAX. Furthermore, error subjects obtained proportionally more disconfirmatory or MED/non-DAX information than did no-error subjects. Finally, there were no differences between informed and uninformed error conditions, whereas, if Tweney and Doherty (1983) were right, those subjects informed that error would occur should have relied more on a confirmatory heuristic.

Doherty and Tweney concluded that the presence of error may have so disrupted performance that it created a kind of "floor effect" that obscured any informed/uninformed differences. Therefore, in a second experiment they focused solely on the single-rule condition and crossed error and no-error conditions with informed and uninformed conditions to produce a two-by-two design. This meant that in addition to the three error conditions used in the previous study, there would now be one group of subjects who were told that 20 percent of their trials would be subject to system-failure error when in fact all the results would be reliable. This condition comes the closest to the possible-error conditions described in chapters 4 and 5; the difference is that Doherty and Tweney's subjects were told that 20 percent error would definitely occur, whereas Gorman's (1986; 1989) subjects were told that there was a 0–20 percent possibility of error. Also, Doherty and Tweney gave their subjects no way to indicate where they thought errors had occurred, e.g., by covering them with strips of paper as in Gorman's possible-error studies.

Subjects in both informed and uninformed error conditions combined solved the rule significantly less often than did subjects in both no-error conditions, and were also significantly less likely to identify the correct rule on their first announcement. Again, error subjects proposed nearly twice as many triples as did no-error subjects and significantly more of the former subjects replicated than the latter, though there were no differences in mean number of replications. The authors made no effort to discuss or keep track of replications-plus-extensions, but the difference in the number of triples suggests that SF subjects probably were employing this heuristic. There were no significant differences in terms of proportion of non-DAX triples predicted or obtained: as in the previous experiment, error did not increase subjects' use of a confirmatory heuristic.

Again, there were no major differences between informed and uninformed conditions. Informed no-error subjects were about half as likely as uninformed no-error subjects to solve the rule on the first announcement, but these differences disappeared by the end of the task. Apparently, informing subjects that there is error does add to the uncertainty initially, but—like possible-error subjects working on Wason's ascending-numbers rule—these informed-error subjects eventually realized that there were no errors.

In conclusion, Doherty and Tweney demonstrated that adding system-failure error to the 2-4-6 task disrupts subjects' attempts to solve the problem and causes them to propose more triples and replicate more. But their studies included no direct comparison between possible-error and experimentally induced actual-error conditions, only the indirect comparison between informed and uninformed error conditions. Their subjects were able to obtain "serendipitous disconfirmations" by asking the experimenter if each rule announcement was correct, which may explain why Doherty and Tweney found so little evidence of hypothesis-perseveration (see 2.4.2). Doherty and Tweney concluded that "Any further error research using this or any other hypothesis-testing tasks should be designed to minimize the use of any source of information other than data collection for hypothesis testing (e.g., permitting only one rule announcement)" (1988:129).

Furthermore, while SF error had a major effect on the 2-4-6 task and on a plant-growth MCPL task, its effect on Kern's artificial universe was surprisingly small. Subjects on the 2-4-6 task did employ the methodological falsificationist's heuristic, even though Doherty and Tweney concluded that, "While there were, in the 2-4-6 research, significantly more repeated observations in the SF error than in the no-error conditions, our judgment is that the subjects made far fewer such test-retest reliability checks than they should have" (1988:201). In other words, subjects might not have employed this heuristic sufficiently. In the Kern artificial universe, subjects were limited in terms of the number of tribbles they could plant, so—as in the possible-error studies where the number of triples subjects could play was limited—this greatly increased the cost of replication and made it less likely that subjects would use it. What we do not know is the extent to which subjects used replication-plus-extension on each task.

Overall, SF error has paradoxical effects, depending on what task is used. On the traditional 2-4-6 task, it does appear that actual SF error ought to have a more disruptive effect than the mere possibility of error would have; Doherty and Tweney found significant performance differences between SF and no-error conditions, whereas Gorman found no performance differences between possible-error and no-error conditions. Furthermore, SF error might cause hypothesis-perseveration if subjects are not permitted to ask the experimenter whether their rule is right.

6.4. COMPARING POSSIBLE AND ACTUAL SF ERROR

While Doherty and Tweney were conducting their research into actual SF error, I was embarked on a similar program of research, and sharing my preliminary results with them. Once again I was focusing on questions raised by my own research, rather than by a careful review of previous literature.

I had discovered under what circumstances the mere possibility of SF error significantly disrupted problem solving. Now I wondered whether actual SF error might have powerful effects in situations where possible error did not. So I followed my usual heuristic of conducting a replication-plus-extension: I went back to Wason's original 2-4-6 task and compared actual, possible, and no-error conditions. To review, in my previous study when the possibility of error was added to Wason's traditional 2-4-6 task, subjects in the possible-error condition followed a replication-plus-extension heuristic, and this heuristic was sufficient to eliminate potential errors (see section 4.5).

It seemed a natural step to add actual error to this design, and compare its effects to possible error and no error. I was not then aware that Doherty and Tweney were doing something very similar, but, fortunately, my study would differ in several important respects:

(1) One of the major recommendations to emerge from the Doherty and Tweney research was that future studies or error "should be designed to minimize the use of any source of information other than data collection for hypothesis testing" (1988:129). This meant that in hypothe-

sis-testing studies, as in MCPL studies, experimenters should not tell subjects whether their guesses about the rule were correct or not. From the beginning, my 2-4-6 studies had included this "no experimenter feedback" feature (see 2.4.2). It was apparent that this feature should be extended to studies of actual SF error.

(2) I had developed a method for subjects to indicate on what triples they thought error had occurred; Doherty and Tweney's studies included no such method.

(3) I was sensitized to the potential role of replication-plus-extension, whereas Doherty and Tweney were not.

(4) I decided that, for the purposes of comparison, every actual-error subject ought to experience the same pattern of errors, to eliminate the possibility that different patterns of error would produce very different responses. Doherty and Tweney, in contrast, had generated a different pattern of errors for each subject, and had noted that SF error had its greatest effect on a small number of subjects. Perhaps these individuals had experienced patterns of error that were different from those of other subjects. The pattern in which errors occurred might be a variable worthy of study in its own right.

These differences meant that my research would have a unique contribution to make. So would Doherty and Tweney's: their 2-4-6 research was part of a much larger and comprehensive program of research, which included comparisons I did not have the resources to make, e.g., between different types of error and different tasks.

6.4.1. PROCEDURES

To explore this new phenomenon, I followed my standard heuristic of going back to a familiar experimental design when I wanted to add a new variable. I used the same procedures as in my first 2-4-6 possible-error study, with three exceptions:

(1) Actual SF error condition: These subjects were given the same instructions as possible-error subjects. But the feedback they received on their triples included 20 percent experimentally induced "actual" error, determined in advance by means of a random numbers table. By chance, the first two triples subjects proposed after the initial 2,4,6 were "errors," which meant that if each was actually correct, it would be classified as incorrect, and vice versa. This pattern of errors, which arose by chance, struck me as particularly interesting: what happens when error is encountered early, before subjects or scientists have had time to formulate hypotheses?

I compared this actual-error condition to the standard possible-error and control conditions, using Michigan Tech students, fifteen of whom (by random assignment) ended up in the actual-error condition, fourteen in the possible-error condition, and thirteen in the control condition.

(2) Instead of the asterisks used to mark potential errors in the original

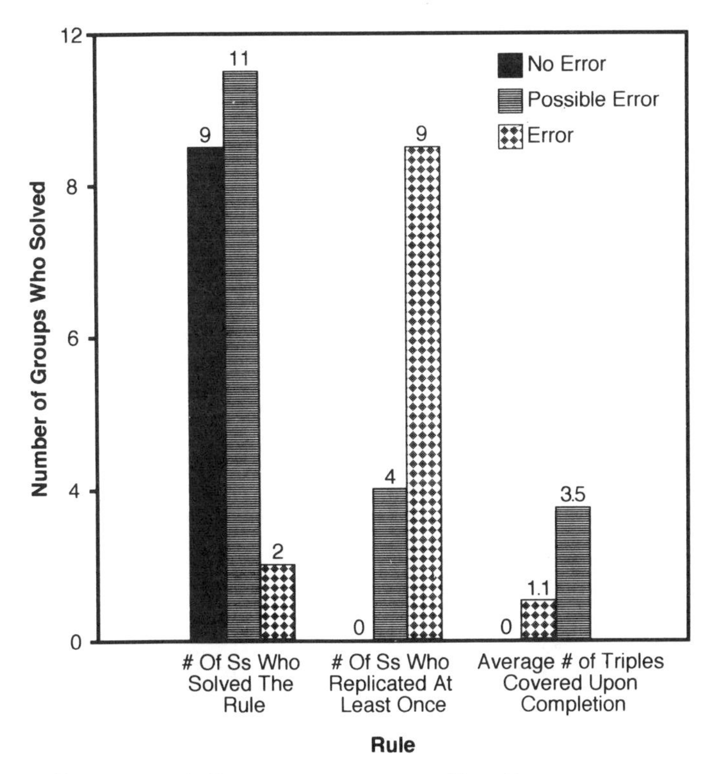

FIGURE 6-1 A COMPARISON BETWEEN NO-, POSSIBLE-, AND ACTUAL-ERROR CONDITIONS ON THE 2-4-6 TASK

study, subjects were given strips of paper to cover them, as in the experiments reported in chapter 5.

(3) Time was included as an additional dependent variable: each subject wrote down his or her starting and finishing times to the nearest minute. Time was added because I wasn't sure that actual error would make subjects propose more triples; instead, it might force subjects to think longer about each triple. My experimental assistants and I made it clear that subjects could take as much time as they wanted.

6.4.2. RESULTS

The differences in number of solutions, number of replications, and number of triples covered were statistically significant. As can be seen from the graph, actual-error subjects solved the rule far less often despite replicating more than did subjects in other conditions. Twelve actual-error subjects recognized that they had encountered error; only four possible-error subjects believed they had.

Actual-error subjects also proposed more triples and took more time to complete the task; this extra effort was not sufficient to permit most of

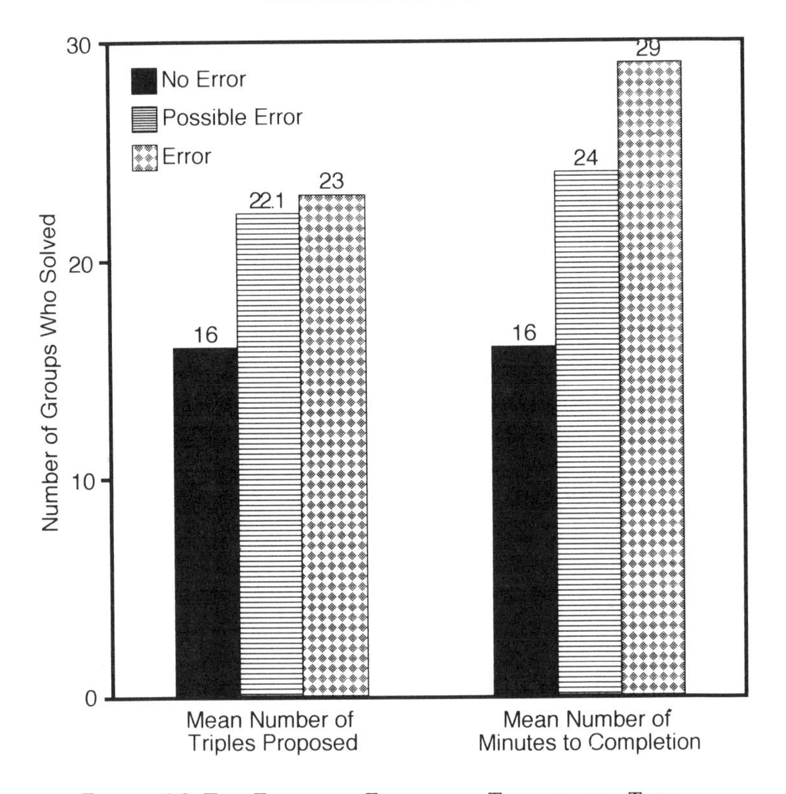

FIGURE 6-2 THE EFFECT OF ERRORS ON TRIPLES AND TIME

them to solve the rule. Statistically speaking, the difference in the number of triples proposed was not significant; the trend, which was in the right direction, was swamped by the amount of individual variation. In other words, individual subjects varied so widely in the number of triples they proposed that any differences between conditions were not significantly greater than the individual variations. The amount of individual variation in time to completion was similarly disparate, but there were even larger differences in the mean times for each condition.

Differences in variation were largest on the replications measure. What this means is that a few individuals replicated a great deal; for example, in the actual-error condition, one subject replicated fourteen times and another ten times. Therefore, the mean differences in replications among conditions must be interpreted cautiously because they are heavily influenced by a small number of subjects. Similarly, Doherty and Tweney found that their results were greatly influenced by a few subjects. However, nine of the fifteen actual-error, four of the fourteen possible-error, and none of the no-error subjects replicated at least once, which confirms that actual-error subjects really were more likely to replicate.

The performance of possible-error and no-error subjects in this study replicated the results reported in the first possible-error study using the 2-4-6 task. The only difference is that possible-error subjects in the earlier study did propose significantly more triples than did control subjects, but clearly the trend in both studies was the same. Once again, possible-error subjects were able, for the most part, to use replication-plus-extension to eliminate the possibility of error and then solve the rule.

6.4.3. Processes of Actual-Error Subjects

Only two actual-error subjects solved the experimenter's rule. Nine showed evidence of hypothesis-perseveration and came up with final hypotheses that explained all the data they had available, but were less general than the target rule: guesses included doubling the first number to get the second, then adding the first and second to get the third; adding two to the first to get the second and four to the second to get the third; adding a constant to each number to get the next; and rules involving ascending multiples.

Two of these actual-error subjects simply followed a confirmatory heuristic for a few triples. One decided, after proposing 4,8,12 and 1,2,3 and finding they were both wrong (because of the error), that the rule must be "n, n+2, n+4" and proposed only four more triples, all confirmatory. The other, after 8,12,16 and 3,5,7 were apparently wrong, tried a rule of the form "n,2n,3n" for twelve more triples and stopped just short of triple number 16, where he would have encountered another randomly induced actual error. Presumably, if they had encountered no errors, these subjects would have simply settled on the "even numbers ascending" pattern suggested by their first triples.

Similarly, one additional subject followed a confirmatory heuristic throughout, but she proposed more triples than these first two and therefore encountered more randomly induced actual errors. She tried "numbers in the triple must go up by two" on the first two and, when both of these were wrong, switched to "the first two numbers of the triple add up to equal the third." She proposed 24 triples in a row that followed this rule, accurately dismissing as errors the only four places it failed to work. In this case, a confirmatory heuristic was a powerful tool for detecting whether and where errors occurred, but in the absence of any systematic attempt to disconfirm, she missed the target rule, which contained her hypothesized rule as a subset. Her confirmatory strategy would have been more successful if the target rule had been contained within her hypothesized rule (cf. Klayman and Ha 1987).

The other six subjects who engaged in hypothesis-perseveration also preferred a confirmatory heuristic, although they also made an effort to disconfirm at least occasionally. But they assigned errors to results inconsistent with their final hypotheses. For example, one subject who decided the rule might be "even numbers in ascending order" followed a confirmatory heuristic for eight triples, then switched to disconfirmatory, proposing 5,8,12 and 7,14,23. Eventually, she covered 5,8,12 and 7,14,23 with yellow strips, because they did not fit her final rule. As she said on her final questionnaire, "I assigned errors to the triples I did because they did not fit my hypothesis."

Four subjects came up with final guesses that were not consistent with at least some of the data they had gathered. For example, one subject guessed the rule was "numbers must increase by the same whole number factor," yet just before she made the guess, she proposed 2,3,9 and—noting that it was correct—replicated it. She did not make it an error, nor did she alter her guess in the face of this disconfirmatory evidence. On her final questionnaire, she said, "once I made a guess I tried to pick triples that corresponded with the guess." In other words, she tried to propose confirmatory triples, and when she did propose one that disconfirmed her hypothesis, she ignored the result. Wason (1960) observed a similar process in a situation where there was no possibility of error: a number of his subjects who displayed a "confirmation bias" not only proposed confirmatory triples but also ignored or could not understand disconfirmatory information when they encountered it. This sort of behavior seems incomprehensible from a Popperian standpoint, but makes more sense from a Lakatosian. As noted earlier (see section 3.9.1), some subjects appear to develop hard-core representations of the rule that are not falsifiable; negative evidence has to be dismissed as due to error, covered by corollaries, or ignored.

One subject simply had no final guess at all; he proposed 55 triples and managed to eliminate all of his hypotheses. He pursued a hypothesis involving even numbers in ascending order for about 25 triples, then switched to odd numbers and found they were also correct in ascending order. So he began to work on the idea of multiples, assigning errors to any nonmultiple triples like 23,33,43 that were correct. But this hypothesis also ran into trouble: too many triples without multiples were correct, and several that fit this pattern appeared to be incorrect because of the SF error. He succeeded in correctly identifying only four out of the eleven errors that occurred in his data, in part because he replicated only twice. Other subjects achieved similar error-identification rates; the average proportion of errors correctly identified in this condition was about .41.

On his final questionnaire, this subject stated his heuristic for identifying errors: "Errors were assigned to [triples] that weren't fitting previous patterns." Contra the methodological falsificationist's dictum, error becomes a way to preserve hypotheses and patterns. Collins (1985) refers to a kind of experimenter's regress that occurs when scientists have to decide which experiments are valid and which are erroneous: one of the best indicators of a "good" experiment is that it achieves the "correct" outcome, but one cannot be sure what a correct outcome is until one has conducted a valid experiment! Therefore, when a subject like this one has a large set of consistent results, he naturally uses them to define what constitutes a valid outcome—even though, logically speaking, he should replicate every one of the results he considers erroneous.

6.5. CONCLUSIONS AND SUGGESTIONS FOR FUTURE RESEARCH

The presence of 20 percent system-failure error made it much more difficult to solve a rule that is solved by most subjects under possible-error conditions. Actual error encouraged subjects to rely on a confirmatory heuristic, which helped some of them identify where errors occurred but made it more difficult to discover the target rule, especially as they tended to use error to immunize their final hypotheses. It is often impossible to distinguish between a confirmatory triple and a replication-plus-extension in the absence of protocols where subjects might state their intentions. Collins (1985) argues that confirmatory power increases as the difference of a new experiment from a previous one increases; by this logic, a replication-plus-extension has less confirming power than a triple which is as different as possible from previous triples while remaining consistent with the hypothesis. Clearly, actual-error subjects were using a combination of confirmation and replication. For example, one subject who guessed the rule was "n, n+2, n+4" proposed six triples, all but two of which went up by twos; in other words, 66 percent of her triples could be viewed as replications-plus-extensions. On her questionnaire she noted that she wanted to get "Ys," the classic indication of a confirmatory heuristic. This combined confirmation/replication-plus-extension heuristic would have pinpointed any further errors, but would never have revealed the target rule.

The effect of system-failure error in this experiment was greater than in Doherty and Tweney's experiments, where the effects of SF error were limited to a few subjects, because subjects were given no experimenter feedback on their tentative hypotheses. This left them free to engage in hypothesis-perseveration, pursuing confirmatory/replication heuristics and/or assigning errors to those triples that did not fit their current hypothesis.

These results pose a problem for the methodological falsificationist, because actual-error subjects replicated over ten times as often as did possible-error subjects. Indirect evidence suggests that they also used more replication-plus-extension. Yet at least fifteen of the twenty-one actual-error subjects still engaged in hypothesis-perseveration. In this experimental situation, the cost of replicating was low; in situations where it would be higher, as in the possible-error experiments where the number of experiments that could be conducted was limited, presumably the "disruptive" effect of actual system-failure error would be greater. But results from the Kern artificial universe, where the number of experiments was limited, suggest that SF error might not have such a disruptive effect in these situations; SF subjects in the Doherty and Tweney version of this task did not replicate systematically, yet the SF error posed problems for only a few subjects.

Clearly, more research needs to be done on the effect of actual SF error in situations where the cost of replication is increased, focusing especially on different types of tasks and rules. Consider the type of rule on which the possibility of error has its maximum effect. When the results of the current experiment depend on where it fits into a sequence of previous ones, presumably actual errors would be harder to locate, in part because replication would be far more difficult. Future research should explore the effects of actual error on these sorts of rules.

Philosophically speaking, the methodological falsificationist may still be right—it may be in principle possible to distinguish between Millikan and Blondlot situations by replicating potential falsifications, provided, as Collins would point out, that one can determine what constitutes a valid replication by some means other than whether it agrees with one's theoretical preconceptions. But even on a relatively simple reasoning task, where it is possible to be precise about what constitutes an exact replication, though not a replication-plus-extension, it appears that the methodological falsificationist's heuristic is very hard to follow.

Before rushing to this conclusion, however, we should take the methodological falsificationist's advice and consider that these results are based on a single experiment. A single result with a small number of subjects could be due to a chance error. Or there might be some peculiarity about the procedures involved. Earlier, I noted that the main difference between this study and Doherty and Tweney's studies lay in the absence of experimenter feedback on hypotheses. There is another difference. All subjects in this study experienced the same sequence of random errors; each of Doherty and Tweney's subjects experienced a different pattern. The first two triples proposed by subjects in this study were classified as errors. It could be that this early occurrence of error has a particularly disruptive effect on problem-solving, though a close look at subjects' processes did not reveal any such effect. The early errors did cause subjects to abandon some of their first ideas about the rule, but most quickly settled on other hypotheses that they proceeded to confirm.

6.6. WHEN ERRORS APPEAR LATER

To replicate and extend this result, I decided to do a follow-up experiment using a randomly generated list of errors in which the first occurrence of an error happened later. This study was run under conditions identical to those in the previous experiment, except that the new list of random numbers dictated that the first error would occur on triple number 6 and the second on triple 13 (with three more errors clustered in the next twelve triples to preserve the 20 percent error rate). In other words, subjects would experience no errors in their first five triples and only two in their first fifteen; this would give them time to formulate hypotheses before encountering any errors.

In addition, it was decided to compare only possible and actual system-failure error conditions, because two previous studies had clearly established that the possibility of error caused subjects to propose more triples or take more time than subjects in a no-error condition, but did not affect their ability to eventually discover the target rule.

6.6.1. Results

Once again, actual system-failure error made it significantly harder to solve Wason's "ascending numbers" rule, though the difference was less dramatic than in the previous study: about 65 percent of the possible-error subjects (eleven of seventeen) solved the rule, as opposed to about 30 percent of the actual-error subjects (six of twenty). Only two possible-error subjects played the same triple more than once, as opposed to eight actual-error subjects. The average number of replications proposed by actual-error subjects was slightly lower in this study (2.4) than in the previous one (3.4), perhaps because error was encountered earlier in the previous study. But the pattern of results remained the same: actual-error subjects replicated significantly more often than did possible-error subjects.

Although on the average, actual-error subjects did propose more triples (24) than did possible-error subjects (20), this difference was not statistically significant, which suggests that the amount of replication-plus-extension was similar. In several sessions, subjects failed to record their starting times, so there was insufficient data to determine whether actual-error subjects took significantly more time than possible-error subjects.

Six of seventeen possible-error subjects still had triples covered with strips at the end, which meant that they thought there was some error in the data; all but one of the twenty actual-error subjects had triples covered with strips at the end. This means that most possible-error subjects recognized that there were no errors and most actual-error subjects realized that there were.

6.6.2. A Closer Look at Subjects' Processes

The pattern of results for both actual- and possible-error conditions is very similar to the results of the previous experiment: significantly fewer subjects solved a simple rule in the actual SF error condition. Therefore, I knew I had achieved a reliable result, even though the two studies were so similar that I could not be sure how well it would generalize beyond Wason's traditional version of the 2-4-6 task. While results from both experiments raise some doubts about the methodological falsificationist's claim that replication is sufficient to distinguish genuine falsifications from errors, the methodological falsificationist could counter that subjects are simply not replicating enough: in both studies, only about half of the actual-error subjects replicated.

Some support for the methodological falsificationist comes from one actual-error subject who replicated nineteen times out of thirty-four triples. He began with a complicated idea—that the rule involved Fibonnaci numbers—but quickly abandoned that in favor of "set of three consecutive even numbers." This hypothesis worked until triple six, at which point he tried three consecutive odd numbers and encountered his first random error. The fact that this triple appeared to be wrong supported his hypothesis, but he persisted in checking for alternate hypotheses, obtaining a correct result for 102,103,106, which he immediately replicated. At this point he concluded the rule was "three numbers, ascending order" but he kept testing and replicating; he tried the same three consecutive odds he had tried earlier, then 1,100,200 on triple 13, at which point he encountered his second random error. He immediately replicated that triple three times; the third time he encountered his third random error, so he kept replicating—a total of nine repetitions of the same triple! He did the same thing with the triple 1,8,7, encountering three errors in the course of eleven repetitions. He correctly identified all actual errors and obtained the correct rule, even though he remained hesitant about his final guess; on his final questionnaire, he said, "I am only moderately sure of my final rule. I feel it may possibly be 'three different numbers in ascending order *with more evens than odds*'."

This subject replicated thoroughly, correctly identifying all errors, and switched between confirmatory and disconfirmatory heuristics effectively, testing alternate hypotheses. Therefore, he was able to discover the target rule. One could argue that this supports the methodological falsificationist position: persistent replications plus disconfirmation are sufficient to discover the target rule. But in this case the process of replication took up so much time and energy that the subject was left feeling he had not eliminated all possible alternate hypotheses; he had low confidence in his final hypothesis. If the cost of replication were raised, as in possible-error studies that limited the number of replications or made it imperative that a sequence of trials be replicated, this problem would be exacerbated. Under ideal conditions, the methodological falsificationist's heuristic is helpful: it is in principle possible to identify actual system-failure errors with persistent replications, provided unlimited replications are possible and no other types of noise or error exist to confound the situation.

Of the five other actual-error subjects who solved the rule, one followed a strategy similar to the subject mentioned above, replicating eleven triples of a total of thirty he proposed and employing a disconfirmatory heuristic to check tentative hypotheses. Two others replicated only once and two did not replicate at all. Three of these subjects took fewer triples to solve the rule than the two persistent replicators noted above. These subjects suggest that replication may not be the most efficient heuristic for discovering a target rule. One of these "successful nonreplicators" was protocoled. (I was able to arrange for protocols of occasional, randomly selected subjects on this study because I had several undergraduates helping me with the experiment.) His protocol reveals that, although he never proposed exactly the same triple twice, he did use replication-plus-extension, proposing similar triples. For example, by triple five he said, "I see a pattern of even numbers but I'm not sure if I know the exact pattern yet." On triple six, 12,16,14 was "correct," reflecting the presence of error; at this point he said, "I'm not sure if they have to be in a certain order or if it's just any even number so I'll check; I'll just put them in a different order to see if the answers are consistent." He proposed 14,12,16, which was a replication-plus-extension. When that triple was incorrect, he said, "There's a problem here. It's between triples 6 and 7. I'll put a yellow tab over 7." He proposed 16,14,12 and when that was incorrect, he covered triple 6 instead.

At this point he adopted the hypothesis that the rule was "ascending evens." He tested it by proposing 10,11,12; when that was correct, he exclaimed, "Ooh! Ok, that shot the theory all to heck. With that result it doesn't have to be every even number. Let's see if we can go odd numbers increasing." He tried 11,13,21, which was "incorrect" because this was one of the triples designated to receive random error in feedback. He immediately recognized that it might be "one of the false answers" and did two more replication-plus-extensions, proposing 23,25,27 and 13,15,17; when both were right, he covered 11,13,21 with a yellow strip. He then tried 1,2,3, which appeared to be wrong because triple 16 was an error. He used replication-plus-extension again, trying 3,4,5 and 9,10,11. Both were right, so he covered 1,2,3 with a yellow strip. He concluded the rule was "any three increasing numbers."

In response to the question about whether error affected his strategy, on the final questionnaire he responded, "Yes. Every answer had to be checked. I assigned errors to triples that did not follow my temporary rule at the time." His heuristic for checking was replication-plus-extension, i.e., he did not propose exactly the same triple twice, but instead proposed triples that followed the same pattern. The advantage to this heuristic is that it allowed him to check for error and confirm the pattern at the same time.

Two of the three other successful actual-error subjects also employed replication-plus-extension, in combination with disconfirmatory and strong-inference heuristics. As one said on his final questionnaire, "I thought of just any increasing even numbers, all numbers corresponding to the first letters of the triple, and odd numbers were not ruled out either, so I tried those. My strategy was to figure out a temporary rule and then challenge that rule keeping in mind the 20% possibility of error."

One final successful actual-error subject solved the rule in eight triples and clearly failed to check some alternate hypotheses; while writing his questionnaire he realized he had failed to try a descending triple. His success, therefore, was due in part to "luck" or "insight": he happened to see the target rule almost immediately. But even he used replication-plus-extension and disconfirmation. He started with the idea that the rule was "ascending evens," proposing two replication-plus-extension triples—2,4,6 and 4,6,8—to make sure there were no errors. Then he tried a potentially disconfirmatory triple—1,2,3—which was correct, forcing him to alter his hypothesis to "ascending numbers." His next triple, 7,8,9, was a replication-plus-extension; the next was a confirmatory triple, 1,5,9. When it was wrong, he proposed 1,3,5, a confirmatory triple, then replicated 1,5,9 and concluded he had solved the rule.

Replication-plus-extension is a more powerful heuristic on this sort of rule than the methodological falsificationist's, because it allows one to combine replication with additional confirmations of a promising hypothesis while leaving open the possibility of serendipitous disconfirmation. Philosophically speaking, replication-plus-extension shouldn't really qualify as replication: if one does an experiment under conditions A and gets a result X, then does an experiment under A but with extension B and gets the same result, one cannot be sure that A gave a valid result; it is possible that A + B would ordinarily lead to a result different than A but that an SF error when A was conducted masked the difference. To put it in terms of Wason's task, it is possible that 2,4,6 will be right and 24,26,28 will be wrong because of the rule, which might, for example, prescribe that numbers over 20 must go up by fours. One way around this is to try a number of different extensions; this will help one check for error and also get information about the target rule. Future research might explore the effectiveness of instructing subjects to employ a replication-plus-extension heuristic on a variety of tasks and situations.

Of the fourteen actual-error subjects who did not solve the rule, eight showed evidence of hypothesis-perseveration, i.e., they maintained the same hypothesis throughout. For the most part, these subjects combined replication-plus-extension with a confirmatory heuristic. Four arrived early at a hypothesis involving numbers going up by twos and assigned errors to triples that did not fit the pattern. For example, one subject proposed four triples that went up by twos, then—on triple six (the error)—he switched to a disconfirmatory heuristic and proposed 2,3,4. Because of the error, that triple was incorrect. He followed with a replication-plus-extension of the negative result—5,7,8—and when that was correct he covered it with a strip of paper, tried 13,15,17 to confirm his hypothesis, and stopped. In a no-error condition, this subject's pattern of perseveration might have been broken because 2,3,4 would have been correct.

Two others began and ended with hypotheses involving multiples of three: one focused on adding the first two numbers to get the third and one focused on raising numbers by powers. All used errors to immunize their hypotheses.

One subject who did not perseverate initially began with a variety of ideas but settled quickly on an ascending pattern. When, due to an error, a descending triple worked, he added descending to his guess and embarked on a degenerating research program. His last eight triples descended, and every one was wrong, but he simply couldn't let go of his "hard-core" representation. His final hypothesis was, "Ascending order always works. Descending order works only if there is a rule involved such as three consecutive descending squares." This corollary was ambiguous enough to allow him to retain the descending feature.

One subject who failed to solve the rule employed a disconfirmatory heuristic. He began by looking for a relationship between odds and evens and replicated on about one-third of his triples. Finally, he decided the rule must be "no two numbers in the triple can be the same" and confirmed it by proposing fourteen triples in a row that he expected would be wrong. When all but the three errors were wrong, he concluded he had solved the rule. Here is a case of a subject who persistently pursued a disconfirmatory heuristic and failed to solve the rule because his hypothesized rule was more general than the target rule.

Four subjects never really settled on a rule: they had confusing and incomplete guesses. One of these followed a disconfirmatory heuristic; he was protocoled and made remarks like, "To find things that are correct doesn't help me, I need to [get] things wrong." He proposed a wide range of different three-number combinations and did obtain a high proportion of incorrect triples: 52 of the 75 triples he proposed were wrong. He even proposed "should be strictly increasing" as one of his early hypotheses. He used replication-plus-extension; as a result, he correctly identified about half of the errors. But the other half, which he failed to identify, threw him off badly; ascending triples weren't always right and ones that went up and down weren't always wrong. In other words, this subject seems to have relied too heavily on a disconfirmatory heuristic; he needed to propose a long string of successive triples he thought would be correct to determine where there were errors and whether a particular hypothesis was promising or not.

Another subject who could not guess the rule said, toward the end of a protocol, "The only thing I see is that usually when they're all odd, you say no, and when they're all even, you say yeah, but that doesn't always go, but I guess there's error involved." So he assigned errors to three all-odd triples that were wrong, but remained uncertain about what happened when one or more numbers were even.

A third subject started out thinking the rule involved multiplying a constant by 1, 2, and 3 to produce the triple; his first five triples apparently confirmed it, but then he ran into inconsistent information due to the presence of the errors. He embarked on a degenerating research program, developing increasingly complex hypotheses, involving multiplication, subtraction, and division of one number from another, so that it was hard for him to identify a clear pattern that ought to be replicated. Again, heuristics depend on representations for their implementation. A fourth subject never really settled on a clear representation of the rule or identified any errors. On her final questionnaire she admitted that "Eventually I resorted to randomization and found no correlating rule."

6.7. CONCLUSIONS

Actual system-failure error does have a more disruptive effect than the mere possibility of error on the 2-4-6 task if subjects are given no feedback on the correctness of their hypotheses. This "disruption" generally takes three forms:

(1) Subjects engage in hypothesis-perseveration, pursuing a combination of confirmation and replication-plus-extension, and therefore fail to disconfirm hypotheses that are embedded within the target rule; or

(2) they make some effort to disconfirm their hypotheses, but immunize disconfirmatory evidence by labeling it error; or

(3) they are thrown into such a state of confusion that they cannot come up with a coherent hypothesis.

The task and rule used in the system-failure error studies reported here were fairly simple, the sort of thing most students from the same science and engineering population solve without difficulty under no-error conditions. The effects of system-failure error are likely to be much greater when the situation is more complex. The crash of a British Midland Boeing 737-400 may be a case in point. One of the engines failed, and the pilots shut one engine off. The result was that the plane shook less. They proceeded confidently to landing, but they kept slowing, so they throttled up—and the one remaining engine failed, leading to a disastrous crash (Parker, *Washington Post*, Wed., 11 Jan. 1989). In fact, the pilots had misinterpreted the source of a system-failure error, perhaps because of a faulty warning system, and shut off the only working engine. They might have tried a sort of replication, shutting off one engine and then the other while they were still at high altitude to see which was really the source of the problem—but they either chose not to or thought this would be too risky, and forty-four people lost their lives.

Of course, to generalize from the 2-4-6 task to the crash of an airplane is perilous—but this example does illustrate that SF error is not merely a phenomenon invented in the laboratory. Future laboratory studies should be focused not only on when and how SF error has its greatest effect, but also on ways of coping with this sort of error.

Experiments should also explore what happens when actual error is introduced on the kind of sequence rule exemplified by the Devonian and corn-genetics situations. The mere possibility of error disrupts performance on this sort of task; presumably, actual error would have an even more disruptive effect. What strategies would permit subjects to cope with it?

This sort of externally valid research needs to be coupled with ecological studies of how scientists and engineers make decisions about system-failure errors, including the standard procedures that are used to cope with this and other types of error. Only by a combination of experimental studies and "following scientists around" can we arrive at a full understanding of the role of error in science.

In the next chapter we will be more explicit about how experimental and field research can be combined, including how experiments might play a role in science policy. As a lead into this discussion, we will turn to my attempts to publish this program of research on error, which raises a set of issues regarding the efficacy of the peer-review system.

SEVEN

A Tale of Two Journals

In the first six chapters, I have sketched the biography of a research program designed to determine whether falsification was an effective heuristic on tasks that simulate scientific reasoning. Gradually the scope broadened to include the role of mental representations and the effect of error. This last line of research on error had been especially difficult. My attempts to publish it produced surprising results, as we will see in this chapter, and forced me to again confront issues in the rhetoric of science (see also sections 2.2.4 and 3.4.3).

7.1. THE HYPOTHETICO-DEDUCTIVE FORMAT

Bazerman (1988) discusses the transition from narrative to more hypothetico-deductive modes of discourse in science. For example, Newton wrote what appeared to be a discovery narrative for the *Philosophical Transactions* concerning his experiments with prisms, but in fact his account does not follow the actual order of experiments as reported in his notebooks; he has "cleaned up" his account to make it more persuasive. Criticisms from Hooke, Huygens, and others forced him to report the experiments as though he had done them in a proper Baconian fashion, carefully deducing alternative explanations for results and eliminating them, one by one. Newton's experimental choices may have been motivated initially by the kind of tension that occurs when each experiment suggests a puzzle that must be resolved by the next one; however, his eventual account of his experiments was shaped by the demands of his audience. His *Opticks* became, in Bazerman's words, "a juggernaut of persuasion."

Holmes (1987) shows how Lavoisier presented his experiments in a kind of discovery narrative which was really designed to persuade. He acted as though he had known where his experiments on Priestley's "dephlogisticated air" would lead from the beginning, whereas in fact he discovered a connection between two apparently unrelated sets of experiments as he was writing. In Lavoisier's case, the act of writing produced a new discovery; however, like Newton, Lavoisier had to make it appear to his audience that his experiments had followed a hypothetico-deductive path. Similarly, Manier (1987) argues that Darwin's persuasive rhetoric masks his actual investigative process.[13]

Like Newton and Lavoisier (but with far more modest results), I followed hunches or intuitions derived from previous experiments in my program of research on error. But, as the next section illustrates, in order to report the results I had to rearrange them to fit the hypothetico-deductive format expected by research journals, attempting to establish both a theoretical rationale and a clear justification for each study. These efforts, in turn, clarified the significance of the research.

7.2. A NEW AUDIENCE

I was concerned that all my research so far had been published in British journals; I wanted to make sure that American psychologists were aware of it. So I sent an account of my possible-error research to *The Journal of Experimental Psychology: Learning, Memory and Cognition*, a highly prestigious American publication. I knew that both Wason and Tweney had major papers rejected by *JEP:LMC*, but I felt I ought to try to reach this important audience again.

My paper was given a thorough and prompt review by five referees and was rejected, basically on four grounds, though there were other comments:

(1) The paper did not follow the classic hypothetico-deductive format. As one reviewer commented, "Why does the reader need four experiments? Taken together, they seem like a group of pilot studies all directed toward the same goal and succeeding in varying degrees. Rhetorically, the four are presented as if they were eliminating successive alternative hypotheses, but substantively, they don't support such a logic."

Like Newton in the earliest stages of his optical research, I was still providing a narrative account of my research, walking the reader through each study as it occurred to me, showing how one result suggested the rest without worrying too much about what overarching hypotheses were being tested. True, I had eliminated the pilot studies described in 5.2.4 and 5.2.5 but that made the study-by-study logic harder to follow. I decided this was primarily a rhetorical problem: I needed to transform a discovery narrative into a hypothetico-deductive narrative in which I made it sound like each subsequent experiment had been motivated by theory.

(2) The question of theory was also raised by the reviewers. My references to philosophers, such as Popper and Lakatos, seemed to puzzle them. In contrast, several recommended that I cite the psychological literature on inductive reasoning. I read this as an audience problem: this readership was less interested in philosophy and the 2-4-6 tradition and more in other experimental work.

(3) The analogy between scientific reasoning and experimental work was weak. One reviewer questioned whether one could use college soph-

omores to make inferences about the reasoning of scientists. This criticism surprised me: so much of the American psychology literature is based on the college sophomore! In fact, many of my subjects were junior and senior science and engineering students, but the reviewer had still raised an issue I needed to address. He or she went on to say, "My criticisms should not be construed as a crusade for ecologically valid, naturalistic experiments. Heaven forbid! It seems provocative to study scientific reasoning in controlled laboratory settings. Such an approach has *many* advantages over naturalistic approaches. However, it seems reasonable to expect that any laboratory paradigm be a decent analogue of what it purports to examine."

This reviewer seemed to understand the potential value of experimental work on scientific reasoning. Again, I felt (perhaps wishfully) that the other objections could be answered by changing my rhetoric, rather than my research. In particular, I needed to explain what specific aspects of scientific reasoning could be simulated by the 2-4-6 task, and explain it in a way that an American experimental audience, not steeped in the literature on falsification, could understand.

(4) Several reviewers remarked that these problems might have been of less consequence had my results been "surprising." I was a bit puzzled by this—it seemed to me that obtaining an effect for the possibility of error in one study and no effect in a similar study was the surprise that had motivated the whole sequence of studies.

Although it was clear from their comments that none of the reviewers accepted the assumptions of the invisible college I was a part of, I found most of their criticisms valuable; after all, the point of this submission was to reach another audience, to connect with other invisible colleges. It was true that I needed to sharpen my hypothetico-deductive rhetoric and create a more explicit theoretical rationale. I could certainly cite the best, recent work on induction, by Holland, Holyoak, Nisbett, and Thagard, and show how Popperian terms like *immunize* could be expressed in their language. I could eliminate some of my comparisons with science and strengthen others, to clarify what aspects of science I intended to simulate. I had at that time also finished running my first study on actual error (described in chapter 6), which might constitute a "surprising" result.

I called the editor and he made it plain that he was especially interested in novel results. What I had given him was essentially a replication (or replication-plus-extension) of my previous work. But the idea of publishing the first results on actual error intrigued him. So he told me he would consider a resubmission.

Note the implication that journals, like magazines and newspapers, want "news"—which means that replications are less likely to be published than replications-plus-extensions, which in turn are less likely to be published than novel results. In the last chapter, we saw how difficult it was to follow the methodological falsificationist's advice when the cost of replication was increased. As a career investment heuristic, the cost of replication may be very high indeed; of the three manuscripts I personally submitted that might have been described as replications, none were accepted and one journal editor refused even to referee one because it was a replication (see 2.2.3). According to Mahoney (1987) replication accounts for only about 5 percent of the articles in social sciences journals. Cicchetti, Conn, and Eron (1991), after a thorough review of the literature on peer-review practices across a wide range of disciplines, concluded that:

> With few exceptions . . . the apparent bias against replication studies is very strong (on the part of both reviewers and editors). . . . A successful strategy has been simply to build the replication study into the first part of the research design, after which the main study would follow. While referees and editors, in our experience, seem willing to accept replication studies embedded in an overall research design, they are quite unwilling to accept them alone. (p. 128)

Similarly, my experience with the editor at *JEP:LMC* suggested that one ought to package a replication with a study that contained new results, in effect creating a kind of replication-plus-extension, in which the replication was the first experiment and the extensions were later ones.

Throughout this possible-error research program, I gambled that a replication-plus-extension heuristic I used to explore a paradox would also pay off in career terms. *JEP:LMC*'s response suggested either that this whole research program had indeed been a bad gamble or that I had misread my audience and employed the wrong rhetoric in presenting it. Naturally, I chose to believe the latter. The referee's reports convinced me that I had written a kind of "discovery draft," whose logic was clear to me but not to others. The editor's response suggested the importance of extending my replications further by adding a novel result. If I wanted to reach an audience of experimental psychologists outside of my invisible college, I had to revise.

7.2.1. A Resubmission

I made a set of changes which I thought addressed most of the referee's criticisms. I used hypothetico-deductive language to derive contrasting predictions from the methodological falsificationist's position and from Kern's (1982) arguments regarding hypothesis-perseveration. I compared Popperian language to that of Holland et al. in *Induction*. I added my new data on actual system-failure error, carefully integrating it with the hypothetico-deductive structure of the piece. I did not, however, drop the comparisons with science, nor did I de-emphasize the fact that these studies were motivated, in part, by a desire to replicate earlier results.

Three months later I received another rejection. One reviewer explicitly objected to the idea that the study was a replication: "First, the paper is written from the perspective of trying to replicate Gorman (1986). But this trivializes the research. Instead of portraying a search for replication, it would be better to pursue an exploration of the conditions that determine the effects of real and possible error." This referee went on to suggest further experiments I might conduct—interesting suggestions, but I was surprised that she or he thought replication was "trivial."

One other referee, who had seen an earlier version of the paper, dismissed this one as "largely unchanged." Clearly, I had misread at least one member of the audience I hoped to reach. Two other referees agreed that I was tackling an interesting problem but recommended that I conduct new experiments to address deficiencies in my research program. These comments suggested that I was faced with more than an audience problem: the rationale behind my whole research program may have been misguided.

At this point, I was faced with distinguishing between two hypotheses about my program of research into the possibility of error: (1) my intuitions about the sequence of studies were fundamentally wrong and I would have to conduct substantial additional research before I could publish anything; (2) I was simply trying to publish this research in a journal whose audience was unsympathetic to this sort of work. To find out which of these hypotheses was correct, I did something I had never done before: I took the manuscript *JEP:LMC* had rejected and shipped it off to the *Quarterly Journal of Experimental Psychology (QJEP)* without making any changes. Wason's original 2-4-6 study had appeared in this journal and so had many subsequent ones, including several of my own. Therefore, I knew the audience would be interested in the sort of work I was doing. If hypothesis (2) were correct, they would accept the paper, probably after some minor revisions; if hypothesis (1) were correct, they would reject it.

I received two referee's reports from *QJEP*. Both raised more questions about methodology than about theory. The editor echoed this methodological focus: his major concern was whether information on subjects' hypotheses and heuristics was obtained primarily from what they wrote during the task, or from a post-hoc questionnaire. Retrospective accounts of problem solving are always somewhat suspect (see Ericsson and Simon 1984). I was able to answer this objection by pointing out that my inferences about hypotheses and heuristics were made primarily from data gathered at the time; questionnaires administered after the task provided only supplementary information.

The referees also suggested dropping an entire experiment and making a variety of editorial revisions in the direction of greater brevity and clarity. In contrast to *JEP:LMC,* the editor at *QJEP* was particularly positive about my comparisons with science. Furthermore, there was no objection to my use of the term *replication.* I was able to make all the recommended changes and the paper was subsequently published.

To summarize, my paper had been rejected twice by one major experimental journal and accepted by another. This is normal practice in science; for example, Myers (1990) describes how two biologists' papers went through a series of rejections, revisions, and resubmissions before they were accepted by journals. I was using a task and literature far more familiar to *QJEP*'s audience; after all, they had published most of the papers that used the 2-4-6 task, including two of my own. *JEP:LMC,* by contrast, had yet to publish a single paper from this tradition.

Did this mean that *JEP:LMC* was prejudiced against the 2-4-6 task? One might as easily argue that *QJEP* was prejudiced in favor of it. The fact is, each journal was addressing a somewhat different audience. There was certainly overlap in their readership; I knew psychologists who read both faithfully. But the 2-4-6 "invisible college" was clearly better represented among *QJEP*'s editors, referees, and readers.

James J. Jenkins, in an interview with Baars (1986:243), makes a similar point about trying to publish his experiments on the role of mediating variables in conditioning:

> For example, the early mediation stuff was not regarded as interesting. It's not that editors and reviewers were against it—they just said it wasn't their kind of psychology. We sent one article to Arthur Mellon at the *Journal of Experimental Psychology,* and he didn't even have it reviewed. He simply sent it back with a note, "This is of no interest to my readers." Lots of journals would simply not deal with us, or the referees would give us endless suggestions, "Why don't you do the following set of studies?" There are exceptions to that—I think that Leo Postman as editor of the *Journal of Verbal Learning and Verbal Behavior* was remarkable in his tolerance for dissident views. . . .
>
> I don't think there is any end to resistance. I mean, if you're pushing the edges of the field, you're always up against some kind of resistance of that sort. You just hope for good editors, and you keep making your arguments and coming back over and over again.

Most journals are oriented toward specific invisible colleges and, one might argue, should serve the interests of their audiences. But this also creates the danger that journals will reflect only existing paradigms, making novel work like Jenkins's hard to publish.

7.2.2. Tilting at Windmills

Garcia (1981), in a famous article entitled "Tilting at the Windmills of Academe," describes the difficulties he encountered when he tried to publish his now-classic studies of bait shyness in rats. His work raised two problems for the behaviorist paradigm or research program that dominated studies of animal learning:

(1) All unconditioned stimuli are not equal. Garcia found that rats readily associated sweet water with sickness produced by X-rays, and a clicking

light with shock, but they could form no association between the light and the sickness, or the sweet water and the shock.

(2) Immediate reinforcement is not necessary for classical conditioning. Indeed, one of Garcia's experiments established that rats can associate a feeling of sickness with sweet water drunk 30 minutes earlier.

Both of these findings were initially rejected by animal learning journals. In contrast, Garcia had published about twenty papers in radiobiological journals without a single rejection. This suggests the difficulties involved in switching from one audience to another. Garcia was able to design replication-plus-extension studies that satisfied some of the same journals, in part because he used procedures more familiar to these audiences and in part because repeated exposure to his novel results overcame "editorial neophobia" (Garcia 1981:151).

For example, a number of the referees insisted that Garcia use a control group. Garcia objected that this was a misleading criterion, but in his replication-plus-extensions he was careful to add experimental conditions that addressed his new audience's potential concerns.

Garcia made the further, interesting claim that control groups are required only for those studies that do not correspond to the prevailing orthodoxy. He cites as examples two classic experiments whose failure to employ adequate controls was exposed by later studies:

(1) Guthrie put cats in a box from which they could escape by rubbing a pole in the center. According to him, they learned this response by trial-and-error, eventually rubbing the pole accidentally and then repeating that exact response the next time they were in the box. This study became the basis for his theory of learning.

But Moore and Stuttard (cited in Garcia 1981) found that cats placed in a similar box rubbed the pole repeatedly even when it did not lead to their escape. Basically, cats like to rub themselves on any surface that is handy. As one of Guthrie's critics noted, "With such a powerful technology, one could, no doubt, teach fish to swim" (quoted in Garcia 1981: 154). This falsification of Guthrie's classic result was criticized heavily and emotionally by referees, but a journal editor overruled them and published it anyway.

This Moore and Stuttard study was a kind of replication, but instead of being a replication-plus-extension, it was a replication *minus* an element of the original design: the contingency that was supposed to motivate behavior. This example suggests that future experimental research might focus on the role of replications that are simpler than the original experiment.

(2) Similarly, one of Skinner's classic experiments involved teaching pigeons to peck a lighted disk at the back of a cage by offering them food. But Brown and Jenkins (cited in Garcia 1981) showed that the pigeon would peck the key even in the absence of any kind of reward: "If grain was signaled, the pigeon pecked at the key as if it were grain, and if

water was forthcoming, the pigeon pumped at the key as if it were water" (p. 155). Again, a replication minus an important element made up for the absence of a control group in the original study.

My own experience with American and British experimental journals suggested that journals are sensitive to particular audiences and invisible colleges. Garcia's experience suggests that journals are also sensitive to paradigms.[14] Obviously, there is an overlap between my experience and Garcia's: he was trying to reach a particular audience of psychologists who were part of a network of behavioristically inclined invisible colleges. But the news he brought was disconfirmatory. If journals are tailored to specific audiences, how can one make sure the audience is willing to hear "bad news"?

Mahoney (1977) conducted a study in which he sent the same manuscript to 75 journal referees, but in varying versions which had positive, negative, or mixed results in terms of the behaviorist perspective taken by the journal. The referees were far more likley to recommend the positive version for publication (see 1.5). This study suggests that journals might prefer confirmatory results. Cicchetti, Conn, and Eron (1991) cite a host of studies that provide additional evidence of journals' bias against negative findings.

Garcia's experience and Mahoney's results suggest that one good investment heuristic (see 5.1) is to pursue research programs that are likely to produce results consistent with the theoretical perspective of a journal's audience. For example, Mahoney (1989) describes how one of his early papers was accepted by a behavioral journal on the condition that he delete all uses of the word *cognitive.*

These examples suggest that peer review in psychology may serve to reinforce, rather than combat, existing paradigmatic notions. From a Popperian standpoint, such prejudices probably ought to be combatted: anything that hinders the publication of a falsification hinders the progress of science. But from a Kuhnian standpoint, normal scientists in an area ought to be skeptical about research that threatens the paradigm. Both Garcia's and Mahoney's works were eventually published; a Kuhnian could argue that both eventually joined the growing group of anomalies that signaled the downfall of behaviorism.

7.2.3. Prestige and Publication

Ceci and Peters (1982) adopted a clever design to explore the effect of another variable on peer review. They sent papers by eminent scholars to the same journals that had originally published them, only they altered the names and institutional affiliations so that the manuscripts were now being submitted by unknowns from what appeared to be low-status institutions. Three of these duplications were detected; ten others were not, and of these duplications were detected; ten others were not, and of these, nine out of ten were unanimously recommended for rejection. These findings suggest that institutional affiliation is another important factor in getting into print. Like Mahoney's, Ceci and Peters's research suggests another investment heuristic: attach yourself to a prestigious institution (or avoid attachment to a low-prestige institution), not only because it will have more resources but also because its name will increase chances for publications and grants.[15]

A Popperian would insist that it is evidence that counts in science; institutional prestige should not determine what results are taken seriously. A Kuhnian might well agree: what does it matter which institution a puzzle is solved at? From the standpoint of these two philosophers, Ceci and Peters's result is a potential indictment of the peer-review system, confirming the sociologists' view that what counts as important scientific evidence is decided by social negotiations.

But a Popperian or Kuhnian could also argue that the Ceci and Peters result merely provides further evidence that psychology is not a science and that a similar result could not have been obtained in physics or biology. There has been no equivalent of the Ceci and Peters study in the "hard sciences"; perhaps there should be. Cicchetti, Conn, and Eron (1991) note that astronomy and astrophysics journals have much lower rejection rates than journals in the social sciences, and that once a manuscript is rejected by one journal it is unlikely to be published elsewhere. In contrast, the situation in medicine is much more like that in the social sciences: high rejection rates, but articles rejected by one journal are frequently published unchanged in another. Furthermore, the same audience preferences noted above appear to exist among biomedical journals, as illustrated by the following case.

7.2.4. The Prion: Particle by Negotiation?

Taubes (1986) presents evidence that a member of one biological research program (in the Lakatosian sense) regularly negatively reviewed his rivals. The controversy concerned the existence of a protein, labeled the "prion" by its "discoverer," Stanley Prusiner, that could apparently multiply without genes. Prusiner parlayed his discovery into a major, multimillion dollar research empire; at the same time, many other scientists were questioning whether the prion really existed. As is so often the case in controversies, the debate centered around error. Prusiner claimed that he was using better methods than his rivals; they, in turn, claimed his faster methods were less accurate. Prusiner's rivals also made persistent attempts to disconfirm his results, coming up with alternative explanations that they thought fit the data better. Soon the journals themselves were caught in the controversy:

> (A)ny substantial peer review in the field ceased with Prusiner's first prion paper. From then on, when Prusiner submitted articles to journals, he recommended they not be refereed by his competitors—and few were. His com-

petitors, in turn, suggested that their work not be refereed by Prusiner. One journal, *Cell,* for whatever reason, published only papers from Prusiner and his colleagues, whereas *Nature,* for instance, published mostly papers from the competition. (Taubes 1986:46).

The bulk of the evidence obtained by other researchers appeared to negate Prusiner's hypothesis, but as of 1986 he had just received a 4 million-dollar award that put him far ahead of the competition. In effect, Prusiner was in a position to find the infectious agent responsible for slow-virus diseases, whether it corresponded with his original notion of prion or not—although, as one of his competitors said, "even if the scrapie agent was a virus and all the data was staring him in the face, my bet would be that he'd never see it" (Taubes 1986:52). Compare this statement with the remarks of a lunar geologist interviewed by Mitroff (1981:170), who described another geologist's commitment:

> X is so committed to the idea that the moon is Q that you could literally take the moon apart piece by piece, ship it back to Earth, reassemble it in X's backyard . . . and X would still continue to believe that the moon is Q. X's belief in Q is unshakeable. He refuses to listen to reason or evidence. I no longer regard him as a scientist. He's so hopped up on the idea of Q that I think he's unbalanced.

Interestingly, Mitroff's research indicated that the three scientists most committed to their hypotheses in the Apollo program were also judged to be among the most outstanding. This example and Prusiner's suggest that confirmation may indeed be more than a bias—it may be a very successful investment heuristic. In the next chapter we will consider how one might investigate experimentally the potential value of confirmation as an investment heuristic.

To summarize, Prusiner developed a powerful series of investment heuristics that include announcing and naming a new discovery (the prion), using the press and the academic journals to promote that discovery, and conducting a long series of confirmatory studies. Note the way in which refereed journals divided pro and con on Prusiner's research; this echoes my experience with journals that are pro and con on the sorts of experiments I do. Mitroff's study did not include a discussion of whether journals that published articles in lunar geology were similarly divided.

7.3. CHANGING SCIENTIFIC DISCOURSE

According to Fuller (1989:135), "the most interesting [normative] interventions will involve disciplining the flow of scientific communication." In this chapter we have seen how publication practices either reinforce or are at variance with the norms of science, depending on your philosophical perspective. Let us review the major findings:

(1) Journals appear to reinforce the hypothetico-deductive format effectively, though this is more a matter of rhetoric than of practice—scientists learn to report their findings as though the studies were conducted in a hypothetico-deductive fashion, regardless of the actual motivation for the studies.

(2) Journals, at least in psychology, appear reluctant to publish replications. More research is needed to determine whether replications are more likely to be published in other sciences.

(3) Journals are not solely objective purveyors of scientific information; instead, they are sensitive to their audiences and may serve as mouthpieces for different invisible colleges or research programs.

But are these publication practices necessarily bad for science? Let us consider each of these findings in order.

7.3.1. The Hypothetico-Deductive Format

Hanson (1958) criticized the constraints of the hypothetico-deductive format, noting that

> "The H-D [hypothetico-deductive] account begins with the hypothesis as given, as Mrs. Beeton's recipes begin with the hare as given. A preliminary instruction in many cookery books, however, reads, 'First, catch your hare.' The H-D account tells us what happens after the physicist has caught his hypothesis; but it might be argued that the ingenuity, tenacity, imagination and conceptual boldness which has *[sic]* marked physics since Galileo shows itself *[sic]* more clearly in hypothesis-catching than in the deductive elaboration of caught hypotheses." (quoted in Tweney, Doherty & Mynatt 1981:307)

One normative implication of Hanson's view is that the scientific literature would be improved by more accurate accounts of how hypotheses were "captured." Rearranging results to follow a theoretical line increases their persuasive power, but a more narrative discourse, in which the author presented each experiment as it actually occurred, might give a more accurate picture of how hypotheses are generated as well as tested. McGuire (1989:223) argues that

> this preliminary thrashing around is quite proper; what is improper is doing it carelessly and then sanitizing the final report by expunging the major information it reveals, in order to conform to the hypothesis-testing myth. . . . [This] myth limits publication primarily to the well-formed final experiment which yields little information beyond demonstrating the almost tautological point that a sufficiently ingenious, persistent, and well-financed scientist can almost always manage finally to come up with some experimental contest in which a hypothesized relationship obtains.[16]

Leary (1980), pursuing a related theme, has called for scientific psychology journals to allow authors to state their values, purposes, and goals explicitly; this presumably would include a discussion of some of the "non-rational" sources of hypotheses.

In a sense, Hanson and Leary are calling for scientists to adopt the reverse of Latour's (1987) "black-boxing" investment heuristic. One of the ways to make a successful scientific claim, Latour argues, is to link it to as many "black boxes" as possible; these black boxes can include specialized techniques or equipment, esoteric mathematical formulas, and earlier results that everyone takes for granted. The net effect is that anyone who wishes to argue with the claim has to unpack these black boxes and show that their contents have been misrepresented. From the standpoint of persuasion, the scientific journal article really should be a "juggernaut" of tightly connected black boxes. In contrast, Hanson, Leary, and other advocates of more narrative forms of discourse (cf. Marks 1989) seem to be calling for a more open science in which authors deliberately make their assumptions more explicit and their claims more accessible.

What is needed, of course, is some way of empirically comparing the normative value of a more open or narrative method of scientific reporting with the traditional hypothetico-deductive format. One method would be to run a giant experiment, using actual scientists, in which the progress made by those scientific communities whose journals encouraged an open, narrative format was compared with the progress made by those communities using a more traditional hypothetico-deductive format. For example, one could take two similar "cutting edge" areas in molecular biology or elementary particle physics and force the journals in one area to require a format that includes a more honest reporting of the discovery process while the other stuck to the hypothetico-deductive approach. One could then compare progress in the two areas, and also "follow the scientists around" to see if changes in rhetoric affected changes in the way research is conducted. Discovery accounts might make the literature too idiosyncratic and more difficult to integrate; on the other hand, they might make it easier to replicate or disconfirm by making both assumptions and procedures more transparent.

Clearly, this sort of giant experimental simulation is impossible; even if one could convince the scientists and journals to become involved, it is hard to imagine controlling or manipulating every potentially relevant variable. But this example illustrates the potential role of experiments in testing normative generalizations: if such a study could be done, and repeated in different areas, it might settle conclusively the normative value of narrative versus hypothetico-deductive journal formats.

Simpler simulations could be done in experimental laboratories. For example, as my colleague Eric Freedman has suggested, one could set up a semester-length simulation of science involving upper-level students working on an artificial universe like the one developed by Mynatt et al. (1978) and arrange it so that research teams competed to have their results published in newsletters.[17] One could compare what happened when some newsletters adopted hypothetico-deductive formats and others encouraged discovery narratives. (For a more detailed discussion of a semester-length simulation focused on a different issue, see the end of chapter 8.)

7.3.2. REPLICATIONS

One of the advantages of hypothetico-deductive format is the way in which it forces clarification of the relationship between current research and the state of the field. I gradually redirected my chain of studies on possible and actual error to distinguishing between two claims: (1) the methodological falsificationist's assumption that replication is sufficient to distinguish genuine falsifications from errors; (2) Kern's observation that SF error causes hypothesis-perseveration. Results do not unambiguously support either claim: a few subjects seem to be able to follow the methodological falsificationist's heuristic with success, though far more engage in hypothesis-perseveration. The ideal heuristic, when the target rule lies outside the hypothesized rule, is to employ a combination of replication-plus-extension and confirmation early in the environment where error may be present, then switch to disconfirmation when one has a well-corroborated hypothesis. The reflexive analysis and field studies presented in the previous chapter suggest that this norm may be at variance with publication practices, at least in psychology: journals typically do not publish replications and appear to prefer studies with confirmatory results.

An interesting follow-up to Mahoney's (1977) research on how journals responded to negative findings would be to see how journal referees responded to the same results written up as a replication or as a more novel finding. This would help us distinguish whether the main problem is the use of the word *replication* or the fact that the results are an actual repetition of earlier results. One could also present results as a replication-plus-extension to see if journals responded more positively to that strategy.

Sociological research may also support the value of replication. Reproducible results are very persuasive: if you can package or black-box a set of results in a way that laboratories all across the country can reproduce for themselves, then their own eyes will persuade them. For example, Gooding (1990) has shown how Faraday built and distributed a small electromagnetic motor that neatly demonstrated the results of a long chain of experiments (see Figure 7-1). Here we have a kind of "black box" that includes a set of instructions which lead the recipient directly and inevitably to Faraday's result.

In C. P. Snow's *The Masters*, a young physicist exults because he has just conducted a successful experiment in which nature is made to "sit up and beg." Faraday's demonstration generator was a way of giving others the power to make nature "sit up and beg" by following Faraday's instructions. It was certainly possible for others to dismantle this simple

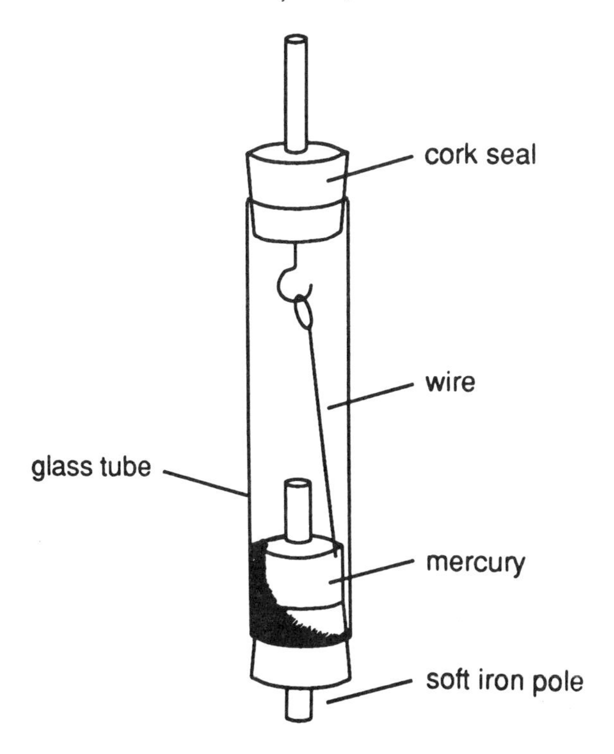

FIGURE 7-1 FARADAY'S DEMONSTRATION DEVICE
Faraday sent this device to other scientists. All a recipient had to do was add mercury and hook up a battery, and the wire would rotate. (From Gooding 1990:190.)

apparatus and explore it—but they could not access the long chain of experiments by which Faraday arrived at this device. Those were effectively hidden, just as a hypothetico-deductive account masks the original discovery process.

So, both experiments and arguments from sociology of science support the importance of replication in science—though, from a sociological standpoint, what constitutes a replication is determined by social negotiations (cf. Collins 1985). More research needs to be done to determine whether psychology journals merely object to the *word* replication, and whether journals in "harder" sciences in fact welcome replications. One could, for example, compare the same data written up with and without the use of the word *replication* and see how it was received by a variety of experimental psychology journals. One would predict that the replication version would be much less likely to be accepted for publication. One could conduct similar studies in other sciences. Failures to replicate ought to be studied as well.

7.3.3. Journals and Invisible Colleges

A third major implication of this chapter is that scientific journals cater to specific audiences representing overlapping invisible colleges. Naturally, more research is needed to determine to what extent this is true in sciences and social sciences. One could, for example, take several prominent journals in different fields and study the links between those who regularly publish in, and review for, them: Who cites whom? What centers, schools, and/or programs are regularly represented? Authors, editors, reviewers, and rejected authors could be interviewed to get a sense of what it takes to publish in these journals. Referee's reports could also be collected and studied. This combination of sociometric and rhetorical analyses could tell us a great deal about the relationship between journals and their audiences.

But if science or social science journals are oriented toward specific invisible colleges, that orientation itself can contribute to a kind of confirmation bias: results that undermine the paradigm or exemplars that form the basis of an invisible college will be less likely to reach its members. One possible alternative is open peer commentary, practiced by such journals as *The Behavioral and Brain Sciences, Current Anthropology, Social Epistemology,* and *Social Studies of Science,* in which controversial articles are accepted and made the focus of a published debate. But even these journals review and reject articles and are sensitive to the needs of their audiences. It would be interesting to see what articles are rejected for open peer commentary.

Whether the notion of journals catering to invisible colleges is a problem depends on one's philosophical perspective. A Popperian would probably find it disturbing but a Kuhnian might argue that journals ought to serve the interests of invisible colleges because rejecting papers that undermine the current exemplars in an area would promote normal science. Which brings us back to the question of how one determines the efficacy of norms proposed by philosophers and others.

One answer is to conduct another giant experiment in which the progress of scientists operating in an environment where the journals are allied with invisible colleges is compared to the progress of those in an environment where each journal is more eclectic. One could do this experiment across a variety of fields—biology, physics, and psychology, for example—to see if disciplinary differences emerge. Once again, this sort of study is hopelessly impractical, but illustrates the potential for using experiments to test the efficacy of normative generalizations.

One could, of course, conduct a more limited simulation along the lines suggested at the end of section 7.3.1. In addition to manipulating newsletter formats, one could see whether subjects made more progress in discovering the rules underlying a complex artificial universe if newsletters were allied with invisible colleges. Such research would have to be complemented by studies of actual scientific practice, such as the sociometric and rhetorical analyses suggested at the beginning of this section. One should adopt a "critical multiplist" perspective, combining a variety of methods (see Shadish 1989 for a good discussion).

One obvious method is historiography. In the next chapter we will consider the question of how historical studies can be used to help assess the strengths and weaknesses of experimental simulations, and how these two approaches might be combined.

EIGHT

From Laboratory to Life

I had a professor in college who claimed that the deeper you went into a narrow scholarly subject, the more you were forced to broaden your scope. This was certainly happening to me. I was applying the very heuristics I was studying, and as I tried to apply them, I was forced to confront the sociological and rhetorical factors influencing confirmation, disconfirmation, and replication.

In the last chapter, we showed how experimental methods, in cooperation with sociometric techniques, rhetorical analyses, and other tools for understanding science, might help us arrive at ideas for improving publication practices. However, we said little about combining these tools with the philosopher's favorite method: analysis of historical examples. If one wishes to make normative generalizations, one cannot limit oneself to current scientific controversies, because it is not clear which journals, theories, and invisible colleges will make the discoveries that last for a hundred years. The advantage of history is that we know the result: whether the discovery was important, whether the theory was accepted. But this knowledge still does not tell us why. Was an idea accepted because of the persuasive rhetoric of its proponents or because they dominated the relevant journals? Or did the winning idea possess some intrinsic "philosophical merit" that made it superior to its rivals?

It is here that psychological methods can play an important role, because they are designed to isolate variables and discriminate which ones play a causative role. The stereotypic psychologist seeks to generalize, to come up with rigorous concepts that will help explain, predict, and improve. The stereotypic historian is loath to generalize and instead tries to immerse herself in the details of a case. I was caught between these perspectives: an experimental psychologist who felt that experiments needed to be closely related to historical cases. Indeed, the more journal referees and my colleagues shook their heads and said my experiments had nothing to say about actual scientific practice, the more I became determined to prove them wrong.

Fortunately, there were exceptions among my colleagues. The philosopher Steve Fuller had pushed me to clarify how experiments might determine the normative value of specific philosophical prescriptions (see 1.3.3, 4.1, and 7.3). But this use of experiments would not help our stereotypic historian, who was more interested in understanding science than in improving it.

W. J. McGuire (1989), a leading experimental social psychologist, suggested that experiments could be a powerful tool for exploring new possibilities and not just a method for testing hypotheses or norms. In addition to determining which historical or sociological variable might have played a causative role, experiments could be used to suggest new ways of looking at historical phenomena. In this chapter we will explore how experiments and historical studies can work together to achieve a richer understanding of technoscientific thinking.

8.1. THE CASE OF MICHAEL FARADAY

As usual, Ryan Tweney (1985) was there ahead of me. He had taken his experimental findings and was using them to provide an enriched perspective on a historical case: Michael Faraday's discovery of electromagnetic induction.

There are parallels between Faraday's heuristics and those of experimental subjects. For example, Faraday combined replication-plus-extension and confirmation early in an environment where he knew error might be present, then he switched to disconfirmation when he had a well-corroborated hypothesis.

In 1831 he began to study how magnetism could induce an electric current. Early on, he used a permanent bar magnet, which he moved in and out of a coil of wire. As he moved the magnet, he changed the strength of the magnetic field surrounding the coil and this in turn generated an electric current in the coil—but only 40 percent of the time. Faraday was not discouraged by this low confirmation rate; instead, he regarded this as evidence that he was pursuing a valid phenomenon.

Faraday worked with two related hypotheses: that a permanent magnet could induce an electric current and that an electromagnet could also induce a current. He switched back and forth between the electromagnet and the permanent magnet in his continued experiments; it is this switching that Tweney saw as analogous to the DAX-MED studies (see section 3.4). For example, Faraday began a series of trials with a more powerful electromagnet; here he obtained confirmation of magnetic induction about 80 percent of the time. Now certain that he had made a reliable discovery and that he could publish his findings, he changed heuristics and tried to disconfirm his hypothesis by systematically testing for artifacts.

Similarly, most experimental subjects used a combination of confirmation and replication-plus-extension in situations where SF error was possible or present, but they had difficulty switching to a disconfirmatory heuristic and, when they did, they tended to assign errors to potential disconfirmations. Faraday knew the importance of using both confirmation and disconfirmation and reflected on it in an 1818 lecture in which he discussed the advantages and disadvantages of "inertia of the mind" (Tweney 1985). Faraday warned of the danger of becoming prematurely attached to one's hypothesis, which might lead to what we (following Kern) have called hypothesis-perseveration. Yet, at the same time, Faraday also recognized that mental inertia had its advantages: it guaranteed that vaguely conceived ideas would not be dropped at the first hint of disconfirmation. Mitroff (1981) has placed a similar emphasis on the value of stubbornness in science (see 7.2.4 and 8.3.2).

The level of error Faraday tolerated in his initial experiments with the magnet was much higher than the 20 percent experienced by the experimental subjects in chapter 6. A confirmation rate as low as 40 percent did not worry Faraday, because he realized that the apparent disconfirmations could be caused by procedural problems. In other words, Faraday was aware that more than one potential source of error was present in his experimental situation. Subjects in chapter 6 experienced only SF error and could only replicate as a means of checking potential errors. Faraday used replication also, but he was comfortable with only a 40 percent rate of replication initially—enough to let him know that an ordinary magnet could produce an electric current. Gooding (1990) has shown how Faraday constantly modified his experimental procedures, in part to make unreliable effects more reliable. In this case, Faraday switched to more powerful magnets. Future experimental research on error must incorporate features which allow subjects to refine their procedures as a way of eliminating potential errors.

8.1.1. Faraday's Mental Representations

Faraday believed in the "unity of forces": "That the attraction of aggregation and chemical affinity are actually the same as the attraction of gravitation and electrical attraction I will not positively affirm but I believe they are . . ." (Faraday, quoted in Tweney 1985:193). Tweney sees this belief manifested in what he calls Faraday's "schema of 'force in general'. Across varying levels of concreteness, the schema incorporates Faraday's thinking from the general (such as his belief in the unity of forces) to the specific (such as his knowledge of how magnetic fields behave in the laboratory" (1985:192).

I introduced the term *mental representation* to describe why the DAX-MED manipulation worked. A schema is a form of mental representation[18] which, as in Faraday's case, helps one to structure one's expectations about what should happen across a wide range of phenomena. Subjects in the DAX-MED condition certainly had different expectations about the rule than did subjects in a control condition. We will discuss the role of another type of representation—mental models—in the cognitive processes of inventors in chapter 10.

Faraday expected the three major forces to be aspects of an underlying unity, and, furthermore, he developed specific expectations about how that unity would be manifested, e.g., his concept of a field. This "unity of forces" idea is what Holton (1973; 1986) refers to as a "thema." Holton does not define "thema" precisely, but he makes it clear that he is talking about fundamental presuppositions that guide the way a scientist approaches problems and studies evidence. These themata appear and reappear over long periods of time; for example, Holton traces the unity thema back to the Ionians, and sees it manifested in the search for a Grand Unified Theory today. In the terminology of cognitive science, a thema resembles a higher-order schema or mental representation that is part of the scientist's sociocultural milieu. Faraday adopted the unity-of-forces thema, but translated it into his own schema about what would happen when he manipulated wires and magnets. Similarly, Millikan's adherence to an atomistic thema or view of the electron led him to expect certain results when he conducted his oil-drop experiments. Note that these terms *thema* and *schema* are frustratingly vague, but do suggest one point where the "social" and the "cognitive" intersect.

Consistent with his representation or schema involving the way in which the three forces could be unified, Faraday hoped to show a form of gravitational induction analogous to the electromagnetic induction he had discovered earlier. To accomplish this, he dropped electrical coils down through hollow, cylindrical towers to see if movement in a gravitational field would produce an electric current in the coil. In his work, he appeared to use a strong-inference heuristic; initially, he obtained confirmatory evidence, so he tried to come up with alternative explanations for the result. He found that there was some magnetic material in the walls of the towers, which explained the slight induced currents he detected. He conducted dozens of experiments on gravitational induction, but always found an artifact to explain the result. In this sense, Faraday was a good Popperian, seeking evidence that would disprove his hypothesis about the relationship between gravity and electricity.

However, Faraday never completely abandoned his hypothesis about gravitational induction, because he knew that all the negative results could be due to experimental errors and that more sophisticated procedures might verify the effect. In Lakatosian terms, Faraday's representation or schema of "force in general" served as the "metaphysical hard core" of his "research program," though Faraday always realized he could be wrong. Like some of the subjects in the error studies reported in chapters 4 through 6, Faraday used the possibility of error to "save" the hard core of his research program from refutation. Is this an irrational strategy? Hardly. Given the procedural difficulties inherent in producing an effect like gravitational induction, no amount of apparent disconfirmations could conclusively prove that some more sophisticated procedure would not detect a subtle effect of this sort.

To put this generalization in different language, Holland et al. (1986)

talk about higher- and lower-order rules. If a lower-order or subordinate rule is falsified, it does not necessarily affect the higher-order rule or representation of which it was a part. In this case, an example of the higher-order rule might be, "each of the three major forces can be translated into the other," and a subordinate rule might be, "if you drop a coil down a tower, its movement through a gravitational field will induce an electric current." Falsification of a subordinate rule does not mean that the higher-order rule is also falsified. Indeed, disconfirmation usually leads to another set of subordinate rules, or what Lakatos would call corollaries. In the case of gravitational induction, Faraday proposed subordinate rules concerning the magnitude of the effect and concluded that perhaps larger coils or greater heights were required to demonstrate the phenomenon.

8.1.2. Representation and Experiment

Faraday's schema guided his experiments; in turn, information from the experiments modified his representations. While it is apparent that some features of Faraday's problem-solving processes resemble those of experimental subjects, Faraday also possessed a wealth of declarative knowledge and an extensive repertoire of procedures (see Gooding 1990 for some examples), and he was working on far more complex and open-ended problems. Gooding (1990) has mapped the way in which Faraday's growing procedural knowledge changed his representations in a chain of experiments using magnets and wires, but he cautions that "Although the analysis necessarily distinguishes between intellectual, practical, material, individual, social, and other strands, I want to emphasize the 'seamlessness' of the web" (Gooding 1989:184).

The experiments conducted so far in this book do not simulate either the declarative or the procedural knowledge of the working scientist; instead, they focus on exploring abstract philosophical prescriptions about how scientific reasoning ought to occur under ideal circumstances. But there is a tradition of experimental research that is relevant to the relationship between domain-specific knowledge, representations, and heuristics on scientific problems.

8.1.3. Experiments on Expert/Novice Differences

In 3.4 we discussed how subjects' mental representations of an abstract task could be altered in a way that dramatically affected their performance. Similarly, Chi, Feltovich, and Glaser (1981) found that experts and novices represent physics problems differently: experts use abstract principles to classify and categorize these problems, whereas novices use surface features. For example, a student who had completed only a single course in mechanics classified two problems that had the word *energy* in their cover stories as conservation of energy problems; in contrast, a physics graduate student recognized that one of these problems involved conservation of linear and angular momentum. Furthermore, the expert's rep-

tion is a good heuristic to follow during the early stages of the inference process, when one is trying to determine whether a hypothesis is worth pursuing. As in Faraday's case, Edison was working in an environment where there were many sources of potential error. As an inventor, he knew that new inventions rarely work the first time and break down as they are tested in increasingly complex environments: a promising invention requires constant refinements. Initial success does not guarantee eventual success, but it was part of Edison's style to forge ahead optimistically, tackling problems that others thought were insoluble (see Josephson 1959 for examples).

As noted earlier, confirmation is also a valuable investment heuristic. Edison knew that the greatest rewards come to the one who creates or discovers first, and that it helps to "blow one's own horn." Anxious to improve his notoriety (in 1875 he had not yet become a household name), Edison concentrated on verifying a revolutionary new hypothesis and worried less about falsifying it. Well aware of the publicity value of a major breakthrough, he announced his new discovery to the newspapers as soon as he felt he had obtained enough confirmatory evidence. (Note that Prusiner seemed to be following an investment heuristic similar to Edison's: name an apparent discovery immediately and announce it to the press.)

While the newspapers heralded Edison's discovery, members of the professional scientific community were more skeptical. In particular, two high school science teachers in Philadelphia, Edwin J. Houston and Elihu Thomson, undertook two sets of experiments to determine whether Edison's etheric force really was a novel phenomenon.

Their first experiments were replications-plus-extensions; in December 1875 Houston and Thomson duplicated Edison's initial findings with an improved apparatus. Although Thomson experienced the thrill of confirming that the sparks could be drawn off metal objects located at some distance from the electromagnet, he and Houston were not convinced that the sparks represented a new force; they concluded that the sparks could be explained by induction.

Edison responded by suggesting that Thomson and Houston did not fully understand the properties of low-resistance electromagnets. Piqued, Thomson and Houston responded in April 1876 by conducting a second series of experiments, designed to disconfirm Edison's claim that the sparks were a new force. Edison had originally claimed that the sparks had no positive or negative charge and therefore were not electrical. Thomson and Houston used two electromagnets which were identical except that one was wound in the direction opposite to the other. Each magnet was capable of producing sparks on its own, but when used together they produced opposing inductive currents and thus canceled out the sparks. Because the two currents canceled each other, it showed that the sparks actually had a positive and a negative charge and thus were not a new force but were instead a form of electricity. Once published, this explanation satisfied most scientists and Edison was forced to abandon his claims to a new force.

Thus, like Faraday and subjects on the 2-4-6 task, Edison began by using a confirmatory strategy in an effort to determine if an apparent discovery was worth pursuing. Unlike Faraday but like many of the experimental subjects, he failed to follow up his discovery by eliminating alternative explanations. Edison represented the sparks as a new force; Thomson and Houston confirmed their existence, but represented them as an aspect of the familiar phenomenon of induction. They were able to disconfirm Edison's claim that the sparks had no positive or negative charge.

So far, the picture that emerges is at least roughly consistent with findings from the laboratory, even though none of the experiments simulated a controversy or any of the investment heuristics that played such an important role in this controversy. One should note the importance of social and rhetorical factors that are not simulated in the experiments. Edison wanted to believe the sparks were caused by a new force in part because he desired the additional fame that would come from discovering it. Thomson and Houston were aware that the scientific community was suspicious of hasty discoveries announced in the newspapers, so they published an article in a scientific journal that verified the sparks but interpreted them in a more conservative way. When Edison attacked their explanation, Thomson and Houston were motivated to prove him wrong. In Kuhnian terms, Edison—an outsider—believed the sparks were an anomaly, evidence for a new force, whereas Thomson and Houston—good "normal scientists"—believed they supported the existing paradigm.

Experiments designed to falsify one research team's hypothesis are often proposed by a rival team, as in the etheric force controversy and the Prion controversy. Edison would never have given up his new force on his own; Thomson and Houston had to falsify Edison's representation of the sparks. We noted earlier that falsifying a rival researcher's hypothesis is a good investment heuristic—especially if the rival is a maverick, operating outside the normal boundaries of the scientific community.

But Thomson and Houston's falsifying experiment disproved only an auxiliary assumption of Edison's etheric force hypothesis, i.e., that the sparks had neither a positive nor a negative charge. In fact, he had stumbled on evidence for the existence of radio waves, though he certainly was not aware of it. Eventually, he came to regret his decision to abandon studies of the "etheric force":

> What has always puzzled me since, is that I did not think of using the results of my experiments on "etheric force" that I made in 1875. I have never been

able to understand how I came to overlook them. If I made use of my own work I should have had long-distance telegraphy. (Josephson 1959:281)

In the etheric force case, Edison might have profited from some of Prusiner's stubbornness, sticking to his hypothesis and adding corollaries when necessary to account for potential disconfirmations. The controversy was one reason Edison developed a contempt for scientists, whom he referred to as "the Bulge-headed fraternity of the Savanic world" (Josephson 1959:279); he was far more interested in practical, working devices.

8.2.1. Conclusions

The Faraday and Edison cases illustrate the way in which experiments can provide a framework for understanding historical examples of scientific reasoning. There is, of course, the danger of falling into a kind of Lakatosian rational reconstruction in which the history is distorted to fit the experiment. Chapter 10 illustrates one way of avoiding this pitfall. In the next section, we will reverse the process, and use a historical controversy to suggest an experiment.

8.3. SIMULATING A SCIENTIFIC CONTROVERSY

It is the theory which determines what we observe. (Einstein)

One of the major philosophical controversies described in chapter 1 concerns the extent to which facts or data are dependent on theories. Obviously, a better understanding of this relationship has important normative implications. In circumstances under which observation is highly dependent on theory, falsification will be at the same time more difficult and more important: more difficult because it would be hard to imagine any observation that could contradict theory and more important because one needs to discover some means of exploring the limits of a hypothesis.

8.3.1. The Modularity Thesis

Fodor (1983) argued that many of the perceptual functions that would be involved in scientific observation, among other things, are modularized, i.e., they are relatively independent and self-contained (see Massaro 1989 for an empirical critique of this view). If so, then the theory would not ordinarily determine what one observes, because perception would be carried on relatively independently of higher-order representations.

De Mey (1989) uses Harvey's discovery of circulation to argue, contra Fodor, that, in science, changes in representation can cause changes in the way observations are processed. He points out that for Harvey an analogy between blood vessels and an inflated glove was very important: "Therefore, the pulsation of the arteries arises from the impulsion of blood from the left ventricle. It is after the same fashion as when one blows into a glove, all the fingers are distended at the same time and mimic the compression of the air" (Harvey, quoted in De Mey 1989:292). This analogy was part of the reason that Harvey interpreted the data available to the Galenists differently, although De Mey is careful to hedge. "It should be clear that there is no single cognitive scheme or model so dominating that only a change at the top is sufficient for producing the conceptual revolution" (De Mey 1989:293). De Mey calls for more research into the interaction between higher- and lower-order cognitive processes in science.

De Mey also leaves open the question of whether Harvey's analogy affected the way he perceived data or interpreted it. Consider the Millikan case reported at the beginning of chapter 6. Millikan interpreted his oil-drop experiments differently from Ehrenhaft, but it is probable that both would have agreed fairly closely on the actual movements of the drops.

Or consider the etheric force controversy. Edison interpreted his results differently from Thomson and Houston, but it is clear that this was not a perceptual problem; indeed, Thomson began by replicating Edison's results.

De Gelder (1989) argues that the theory/observation distinction does not map onto Fodor's modularity thesis in a simple and direct way. In particular, this perception/interpretation issue is ambiguous in philosophy.

8.3.2. The Case of the Canals on Mars

One way to sort through this confusion is to simulate a scientific controversy in which theory appeared to play an important role in perception. Note that one cannot use such a simulation to make a claim like "Theories always affect perceptions"; instead, one may be able to make claims about when and under what circumstances perceptions are likely to be colored by theory.

A controversy in which perception stood at the center was the debate over the existence of canals on Mars. Schiaparelli described "channels" *(canali)* on the surface of Mars in 1877. Earlier observers had detected some vague, streak-like features, but Schiaparelli's "were finer, sharper and more systematically arranged on the surface" (Hoyt 1976:6). Percival Lowell, a wealthy American who developed a strong interest in astronomy, transformed these channels into canals and used them to make a strong argument for intelligent life on Mars. Other observers disagreed, and a heated debate lasting over a quarter of a century was carried on both in scientific journals and in the popular press. This scientific contro-

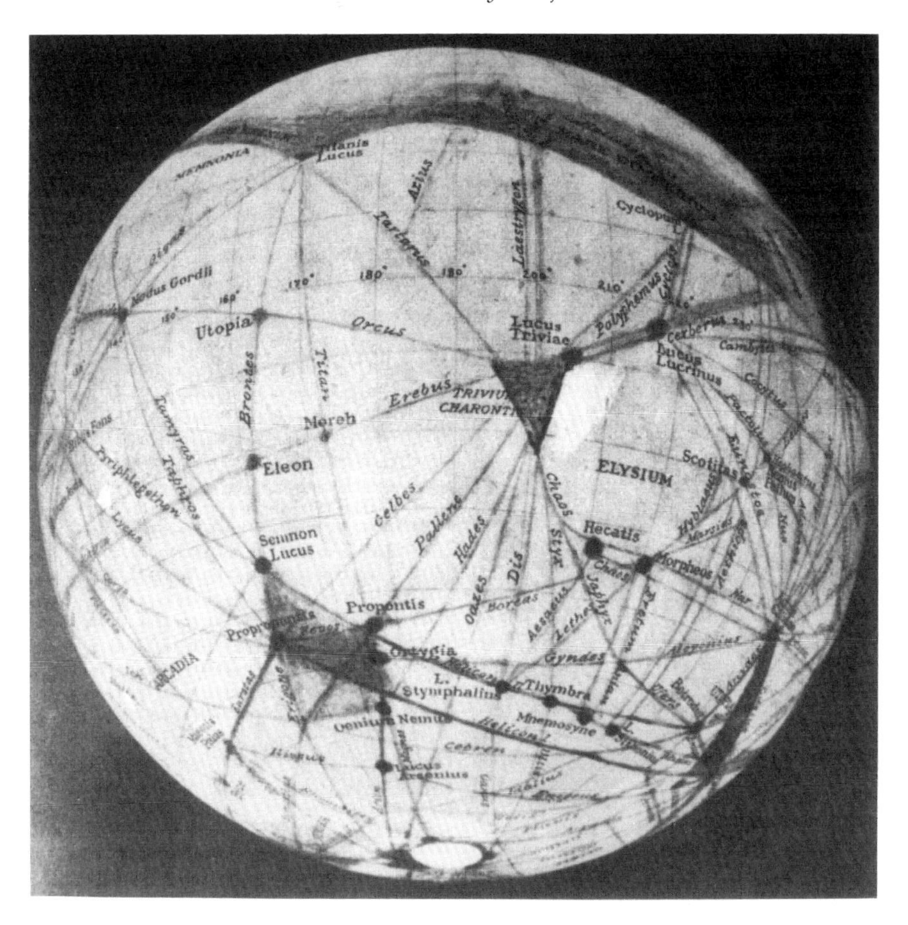

FIGURE 8-1 LOWELL'S DRAWING OF THE MARTIAN CANALS
A globe of Mars prepared by Percival Lowell, showing his canals. (From Hoyt 1976:81; reprinted by permission of the University of Arizona Press.)

versy raises interesting questions about the relationship between theory and data, because some astronomers saw the canals and others did not.

To begin with, let us analyze this controversy in terms of concepts we have used throughout this book. Percival Lowell displayed a strong "confirmation bias": "once his 'general theory' [of the Martian canals] was framed, neither subsequent observation nor mounting criticism during the ensuing Mars furor caused him to abandon or alter any of its essential points" (Hoyt 1976:69). Note that, as Mitroff (1981) has shown in his study of Apollo moon scientists, this kind of stubbornness is often the hallmark of successful scientists:

> The three scientists most often perceived by their peers as most committed to their hypotheses . . . were also judged to be among the most outstanding

> scientists in the program. They were simultaneously judged to be the most creative and the most resistant to change. The aggregate judgement was that they were "the most creative" for their continual creation of "bold, provocative, stimulating, suggestive, speculative hypotheses" and "the most resistant to change" for "their pronounced ability to hang onto their ideas and defend them with all their might to theirs and everyone else's death." (p. 171)

If successful scientists are speculative and stubborn, Lowell certainly should have been successful. In fact, his views were widely circulated and dominated at least popular thinking about Mars for over a decade, though a number of other astronomers remained skeptical. Lowell's hypothesis was that Mars was a dying planet, slowly drying out. The intelligent inhabitants, in order to survive, had built huge networks of canals to irrigate their world, using the water that melted from the polar caps. These canals were surrounded by belts of vegetation, which is why they were visible from Earth.

Lowell was a persuasive speaker and writer, which meant his theory was popular with a wide audience of laypersons. But he also persuaded a number of other astronomers, who independently confirmed the existence of the canals. As more powerful telescopes became available, more and more astronomers agreed that there were no canals. Still, as late as 1924 at least one astronomer claimed to see canals on Mars (Hoyt 1976:170). Lowell died in 1916, confident that he had confirmed the existence of canals and of intelligent life on Mars.

In 1896, Lowell also claimed to see lines on Venus, which he thought were "purely natural" and did not represent canals like the lines on Mars. While Lowell was not the only astronomer to see lines on Mars, no one else ever saw his lines on Venus. He therefore preferred to remain quiet about them, but claimed privately to have seen them again many times. He also played an important role in establishing the rotation of Venus.

Lowell's case seems somewhat analogous to that of Blondlot and the N-ray, except that in the Mars case Lowell was not the first to see "*canali*" on Mars, and other, independent observers disagreed over their existence. Furthermore, photographic evidence was ambiguous, at best. Here we have a phenomenon that was neither clearly replicated nor disconfirmed.

In terms of the error experiments reported in chapters 5–7, Lowell's canals are not simply produced by system-failure error, nor do they correspond to the sort of "deviations from the mean" represented by measurement error; instead, they seem to result from a kind of perceptual ambiguity not precisely simulated by these error experiments. Subjects working on a task like Eleusis could clearly see and identify the card; the question was whether the feedback on the card was erroneous. A situation closer to Lowell's would be one in which the card itself was difficult

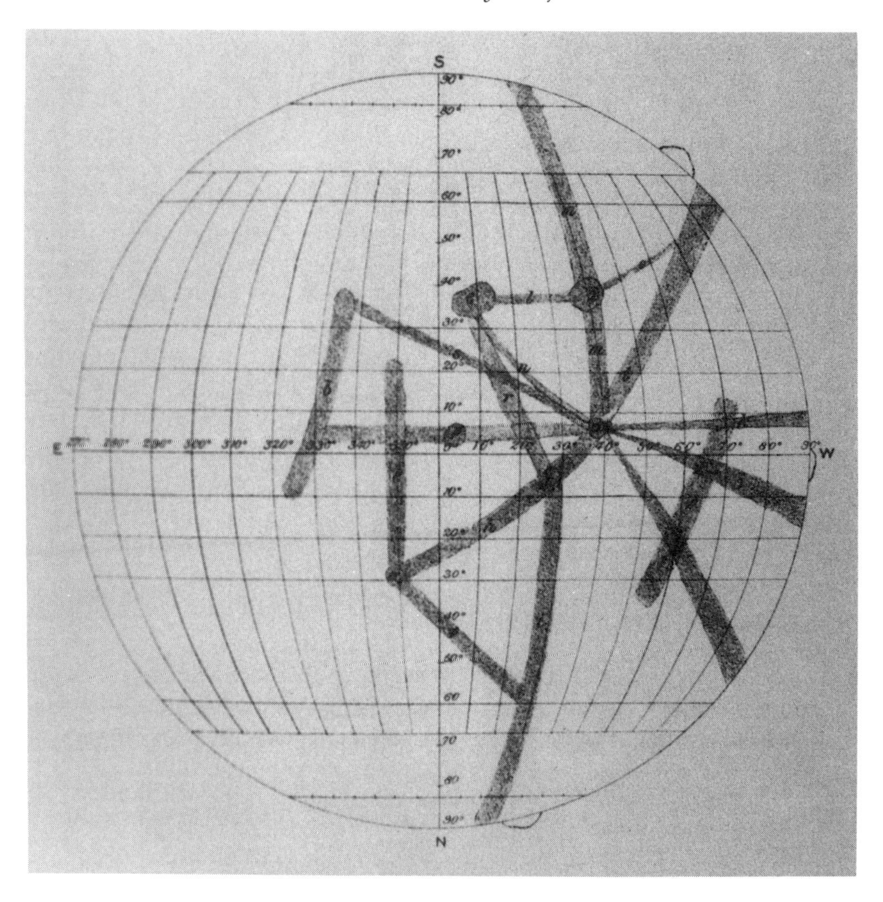

FIGURE 8-2 LOWELL'S SKETCH OF VENUS
A chart showing Lowell's canals on Venus. (From Hoyt 1976:115; reprinted by permission of the University of Arizona Press.)

to discriminate and the feedback might also be erroneous. In this case, the methodological falsificationist's dictum is even harder to maintain: there is a disagreement about the nature of the evidence itself.

8.3.3. EARLY MARTIAN EXPERIMENTS

One of the most interesting things about this controversy is that several attempts were made to resolve it by conducting perception experiments. Douglass, one of Lowell's team of observers at Flagstaff, decided to find out whether Lowell's observations could be caused by some psychological phenomenon: "I have made some experiments myself bearing on these questions by means of artificial planets which I have placed at a distance of nearly a mile from the telescope and observed as if they were really planets. I found at once that some well known planetary appearances could, in part at least, be regarded as very doubtful . . ." (Hoyt 1976:124). Douglass's appearance of doubt caused Lowell to fire him.

Another astronomer, E. Walter Maunder, asked a group of boys from the Royal Hospital School at Greenwich to copy a picture of Mars; even though there were no lines on the original picture, the boys drew them in. Maunder concluded that the lines drawn by astronomers were no more real than those drawn by the boys: "It seems a thousand pities that all those magnificent theories of human habitation, canal construction, planetary crystallisation and the like are based upon lines which our experiments compel us to declare non-existent" (Hoyt 1976:165). Flammarion ran a similar set of tests on a group of French school boys, none of whom drew imaginary lines, but Maunder's critique caused the British Astronomical Association to reverse its position: "The members . . . have themselves seen more than a hundred of the so-called canals during their observations since 1900, yet E. M. Antoniadi, their director and editor, apparently has their concurrence in holding that their eyes were probably deceived and that they really saw something very different from the straight lines they imagined they were looking at" (Hoyt 1976:165).

These early experiments should be replicated and extended. A modern follow-up study might be able to address the issue of whether the canals could have been a kind of perceptual illusion reinforced by the testimony of others and by the fact that Lowell was able to integrate them into a colorful, persuasive theory. If so, this case would provide support for the argument that theories can affect what we perceive.

8.3.4. A Replication-Plus-Extension

One could simply replicate these earlier experiments, using planetary disks. But the average person's knowledge of the planets is much greater now, including the fact that there is almost certainly no other life in the solar system, and might influence how he or she approaches any stimulus resembling a planetary disk. Therefore, one would have to construct a more general situation in which people are shown blurred pictures of terrain features and asked to map them. In one condition, these subjects might be given a detailed, persuasive story about the presence of roads or rivers or other terrain features in this area and be asked to try to include them in their maps. Other subjects would be given no such story. If subjects in the first condition tended to draw the roads in, then one could make the externally valid claim that a persuasive story could affect how people map terrain features. Note that one cannot make the ecologically valid claim that Lowell did this, but it certainly raises the possibility.

Shrager claims that "(1) reinterpretation plays a central role in theory formation, at least in the domains that I have studied; that (2) sensory content is a fundamental and inextricable aspect of the theories formed and their formation by reinterpretation; and that (3) commonsense perception (or something very much like it) is a central psychological mechanism, that plays a key role in both theory formation and complex reasoning'' (1990:438). One could use this canal-controversy simulation to explore the role of perception in theory formation and evaluation, if one added features that allowed subjects to scan terrain and to ''zoom in'' to focus on certain features. What aspects of the terrain would they focus on? How would their new perceptual knowledge affect their overall maps? ''Common sense'', in particular, is at least partly a result of what we learn from others; therefore, this simulation could be expanded to incorporate sociological variables.

8.3.5. Simulating Knowledge Transmission

Consider the classic experiment by Jacobs and Campbell (1961), in which subjects were exposed to the autokinetic effect in small groups. The autokinetic effect occurs when a point of light is shone at the front of an otherwise blackened room; individuals will see the light move differently, depending on each individual's interpretation of his or her own eye movements. When subjects are run as a group, they influence each others' judgments, forming a kind of ''group norm'' regarding the distance the light appears to move (see Sherif 1936). Jacobs and Campbell established that this norm was passed along to new group members as the original group was gradually replaced, member-by-member.

Similarly, one could take individuals who had already mapped roads onto an ambiguous map, and put them together as a group to compare their maps. After they had reached some kind of consensus, one could gradually replace the original members with new members, providing additional opportunities to study similar ambiguous maps. In this manner one could explore some of the mechanisms by which scientific ''exemplars'' are transmitted. One could also encourage a debate between groups that claimed to see roads and groups that did not, and study the way in which the groups sought and interpreted new evidence. One could even investigate which of two competing groups or ''research programs'' was most convincing to outsiders, under a variety of circumstances.

Moscovici (1974) demonstrated that a minority of subjects who adopt a consistent, determined behavioral style affected the perceptual judgments of a majority on an ambiguous perceptual stimulus. Moscovici's research raises an intriguing question: how important is this sort of consistent, determined rhetorical or behavioral style in science? Hilgard, in an interview (Baars 1986), cited Skinner's unwavering commitment to his brand of behaviorism as one reason for its great influence. Similarly, Lowell adopted an unwavering perspective on Mars, for which he argued determinedly and persuasively. How effective was this argument in keeping the Mars debate alive? One could simulate its impact by using a Moscovici design in which a minority argues persuasively for the presence of roads on a map and attempts to persuade the rest of a group. One could couple this to the Jacobs and Campbell design to explore whether the minority's influence extended across "generations."

The point is, designs of this sort permit us to study the interaction of social and cognitive factors on tasks that simulate a scientific controversy. Independent variables could be derived from the philosophical literature, as was done in the early studies of confirmation and disconfirmation. Some sociological variables could also be manipulated, though laboratory tasks of relatively short duration do not provide a good simulation of either investment heuristics or the development of invisible colleges.

One might get around this by designing a very complex environment, similar to Mynatt et al.'s (1978) artificial universe, and allowing students to spend a semester or even a year interacting with it (see section 7.3.3 and note 17). My alma mater, Occidental College, offered a history course in which the students adopted roles of major figures in European history as of 1905. In effect, they learned history through simulating it. Perhaps one could build a similar course around an artificial universe and have students work in competing labs, publishing in journals and struggling to obtain grants. The danger in a study of this complexity is that one could, with the same amount of time and effort, study interactions between scientific groups. The only advantage of an experiment of this sort lies in the possibility for manipulating variables such as behavioral style, investment heuristics, and error. In any case, an experimental simulation of the observational problem faced by Lowell would need to be complemented by a detailed study of historical and sociological evidence. How did Lowell form his initial mental map of Mars? Did he derive his theory from his preliminary observations, or was he already inclined to believe in canals on Mars? How similar were the maps produced by different astronomers, and how often were such maps exchanged? In chapter 10 we will use concepts derived in part from laboratory studies to develop a framework for studying the cognitive processes of scientists and inventors.

8.4. CONCLUSIONS

In this chapter, I have tried to show how laboratory simulations can complement historical studies, each method providing the other with new ideas and perspectives. This combination could be used to achieve such goals as clarifying the role of theory in observation. Moreover, this history/psychology collaboration inevitably leads to the consideration of issues dear to sociologists of scientific knowledge.

The research proposed at the end of this chapter brings us full circle, back to the problem of simulating the social aspects of epistemology (see 1.5). This chapter should have demonstrated that an experimental approach can be adapted to group as well as to individual analyses of science (see Shadish, in press, for a more complete discussion). But experiment is not the only method that can be used to simulate science. Indeed, a great deal of recent attention has been focused on another kind of simulation. That is the topic of the next chapter.

NINE

Using Technology to Study Technoscience

Latour (1987) has coined the term *technoscience* to indicate the way in which science and technology are inextricably linked. In the preceding chapters we have explored the implications of using a particular scientific method to understand technoscientific thinking. Recently, others have taken a different approach, applying a new technology—computers—to the study of scientific discovery and hypothesis evaluation. These computer simulations will allow us to revisit issues such as confirmation bias and external validity, especially because the programmers of these new simulations make claims bolder than those of the experimental researchers we have been studying—claims I initially found outrageous, given my conservative experimental background. I had been trained to make only the narrowest, most circumscribed generalizations from my research, and here were cognitive scientists like Herbert Simon—one of my heroes—claiming that their computer programs "discovered" (see 9.2 below). Indeed, they argued that their programs represented a new theoretical language for cognitive psychology, one that would sweep all before it.

My initial outrage was somewhat hypocritical, of course. I was trying to take my favored method—experiment—and push it as far as possible, to see what its limits were. In a sense, Simon was trying to do the same thing. In this chapter we will explore whether these computational methods represent a new "strong program" that threatens to dominate technoscience studies, or whether they can be used in collaboration with experimental, historical, and sociological approaches.

9.1. MUST COGNITIVE SCIENCE BE DONE ON A COMPUTER?

Johnson-Laird (1988:52) reflects a view common among cognitive scientists when he argues that

> theories of the mind should be expressed in a form that can be modelled in a computer program. A theory may fail to satisfy this criterion for several reasons: it may be radically incomplete; it may rely on a process that is not

> computable; it may be inconsistent, incoherent, or, like a mystical doctrine, take so much for granted that it is understood only by its adherents. . . . Students of the mind do not always know that they do not know what they are talking about. The surest way is to devise a computer program that models the theory. A working computer model places a minimal reliance on intuition: the theory it embodies may be false, but at least it is coherent, and does not assume too much. Computer programs model the interactions of fundamental particles, the mechanisms of molecular biology, and the economy of the country.

One caveat is that such simulations can give radically different results, depending on the assumptions under which they are set up.[19]

Baars (1986:181) goes a step farther than Johnson-Laird goes: "computer programs provide a natural theoretical language for stating psychological hypotheses in a very explicit way. And any theory that cannot be made explicit enough to run on a computer is ipso facto a faulty theory." Now all noncomputational theories are excluded in a manner reminiscent of Popper's attempt to exclude nonfalsifiable theories. For Pylyshyn (1979), computer programs will play a role in the cognitive sciences similar to the role played by geometry in physics: just as every physical theory must be mathematized now, so every cognitive theory will have to be computerized. Those who are skeptical about such a constraint are cast in the role of Aristotelian scholastics, who resisted the mathematization of science (see Fuller 1989 for a discussion).

9.1.1. Computational Models and the 2-4-6 Task

Both Johnson-Laird and Baars would argue that the results described earlier in this book ought to be expressed in the form of a computer program. Let us consider one crude example of how this might be done. In section 3.7 we discussed how confirmation could be an effective heuristic in the pursuit phase of the inference process, but that scientist or subject would have to switch to a disconfirmatory heuristic when trying to justify a hypothesis that had succeeded in producing a string of confirmations. Figure 9-1 shows how this process might be translated into an extremely simple flowchart illustrating the "confirm early, disconfirm late" heuristic.

This flowchart is normative in that it does not purport to describe the actual performance of most subjects; instead, it tries to specify the optimal combination of confirmation and disconfirmation. To provide a computational test for this model, one would want to implement it in a computer language and see if it solved 2-4-6 problems efficiently. If it did, one could claim that the model had achieved a certain kind of external validity, somewhat different from that provided by experiments. From an experiment, one can derive externally valid claims about human performance under ideal circumstances; from a computer simulation, one can derive claims about the ideal functioning of heuristics.

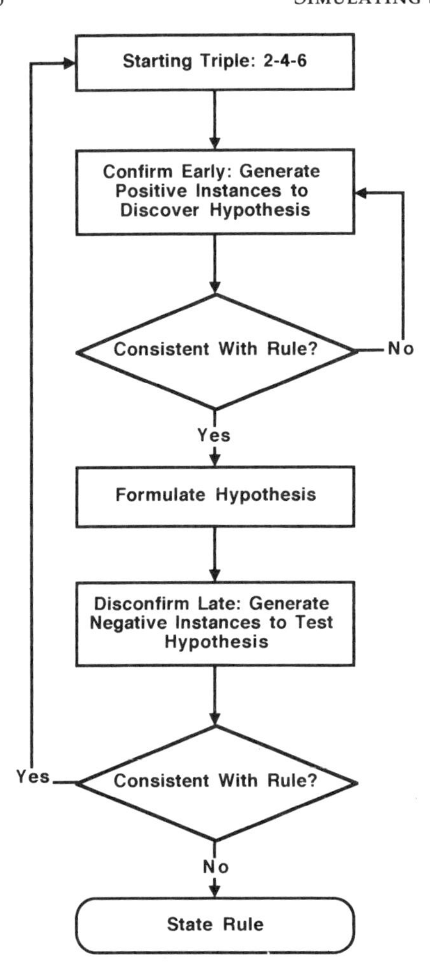

FIGURE 9-1 A FLOWCHART MODEL OF THE "CONFIRM EARLY, DISCONFIRM LATE" HEURISTIC ON THE 2-4-6 TASK

Like experiments, computer simulations can be expanded to model situations that are more ecologically valid. For example, one could add replication heuristics to cope with error to our computational model of the early/late heuristic (see Figure 9-2). One could go a step farther and try this heuristic on the sort of data actually considered by scientists. In a moment we will consider work of this sort by Langley, Simon, and others.

What computers cannot tell us is how and whether human beings actually employ such heuristics. Computerized reconstructions of human problem-solving have to be closely linked to detailed experimental and/or ecological studies.

Consider the 2-4-6 flowcharts in Figures 9-1 and 9-2. Taking a computational approach forces us to be more rigorous; the diamonds in Figure 9-2 for "Replicate" and "Replicate-Plus-Extend" would have to be translated into precise sets of functions in order to be implemented in

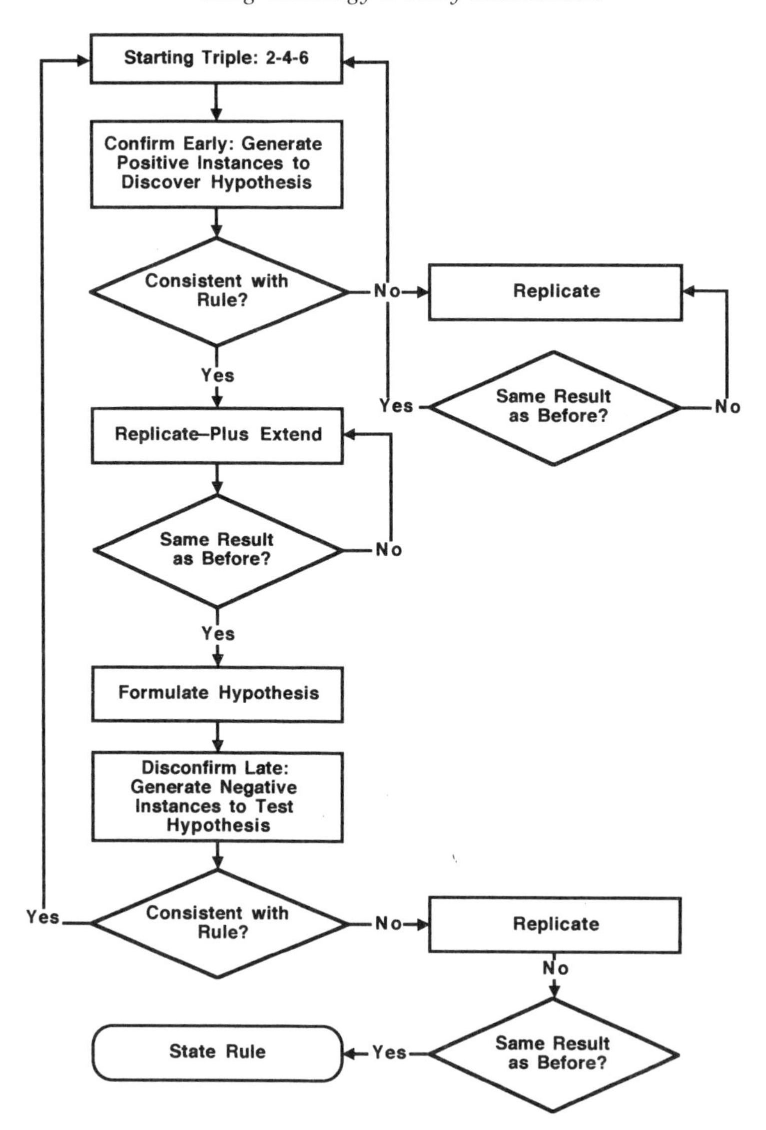

FIGURE 9-2 ADDING REPLICATION TO THE "EARLY/LATE" HEURISTIC

computer code. But further research would have to determine whether this additional precision is useful in describing human processes. Can a computer simulation reproduce the kinds of representations human beings form when they employ these heuristics?

Baars, Johnson-Laird, and others imply that all psychological studies ought to be conducted in a way that is amenable to computer simulation. This constraint could lead to a kind of confirmation bias in which one selects only experimental situations that are amenable to computation. Behaviorists made a similar error when they looked only at the behavior of animals in very restricted settings.

9.1.2. Resisting the Computational Constraint

Several major participants in the development of the cognitive sciences do not accept the constraint that a cognitive theory must be implementable on a computer. In a discussion of the computer program as theoretical language issue, George A. Miller, "the single most effective leader in the emergence of cognitive psychology" (Baars 1986:198), remarked:

> "I really began with the assumption that scientific psychology was possible, and that you could put order into the field in the same way physicists put order into their field—by mathematics. . . . Then I learned that anything you can do with an equation you can do on a computer, plus a lot more. So computer programs look like the language in which you should formulate your theory. But the farther I go with that, the more limitations I see.
>
> "There are things that I deeply believe are true which I don't know how to capture in programs. It isn't so much a matter of finding that mathematics is wrong, or that computer programs are wrong—none of them seems adequate to say what we've got to say." (George A. Miller, quoted in Baars 1986:218)

Ulric Neisser, whose *Cognitive Psychology* (1967) crystallized the cognitive revolution, refers specifically to computational approaches when he argues that

> "It's not necessarily a great step forward to have a theoretical language. The question is what is being said. A step forward would be discovering things that we didn't already know. Finding a new language and using it can be such an exhilarating experience that people don't bother to find out anything with it. I think it may be best to stick close to our intuitions about real situations. One of the important discoveries of cognitive psychology, for example, is Eleanor Rosch's finding that many everyday concepts are organized around prototypes. . . . She came to the prototype idea by considering natural concepts—what do people mean by 'dog', or 'chair', or 'table', or 'furniture'? It turns out that those concepts don't have the sort of structure that had been described in the theoretical language of AI, or the very similar language of earlier concept-formation theorists. Such discoveries will be made over and over again if we look at how things really are." (Ulric Neisser, quoted in Baars 1986:283)

The danger is that psychology could be forced into a kind of computational circularity, in which programs provide effective models of human cognition because cognition must be described in computational language. This argument is reminiscent of the old behaviorist definition of a reinforcer as anything that increases a response, a claim that could not be falsified by definition.

While I am uncomfortable with the idea of forcing psychological theories into a computational Procrustean bed, I would be just as uncomfortable with any attempt to prescribe universal adoption of the experimental method. Computational simulations, like experimental ones, have an important role to play in psychological studies of science and technology. To explore that role, we need to consider examples of the simulations.

9.2. WILL THE NEXT KEPLER BE A COMPUTER?

Langley, Simon, Bradshaw, and Zytkow (1987) describe a series of computer programs designed to discover scientific laws, using heuristics to induce rules from data. These computational simulations are an extension of Simon's extensive program of research on human problem-solving (cf. Newell and Simon 1972). His goal in the computational simulations is to show that scientific discoveries are made by applying the same sort of heuristics used in other problem-solving situations.

Langley, Simon, Bradshaw, and Zytkow named their first series of programs BACON in honor of the most famous advocate of induction. They argue that philosophers like Popper, Lakatos, and Kuhn have spent too much time arguing about justification and too little time considering maxims for discovery: "The efficacy ('rationality,' 'logicality') of the discovery or generation process is as susceptible to evaluation and criticism as is the process of verification" (Langley et al. 1987:39). In fact, they reject the discovery/justification distinction:

> When a search process is provided with heuristics, especially when those heuristics take the form of tests of progress, the partial results of a search receive confirmation or refutation, or some evaluative feedback, before a full-blown hypothesis is produced. Each step or group of steps of the search is evaluated in terms of the evidence of progress it has produced, and the continuing search process is modified on the basis of the outcome of these evaluations. The confirmations of partial results cumulate and make the confirmation of the final hypothesis coincide with its generation. In particular, an inductive generalization is confirmed in the very same way it is generated. (Langley et al. 1987:57)

Note that this search process resembles the "confirm early, disconfirm late" heuristic discussed earlier (see 3.7). But Langley et al. would presumably reject this early/late distinction, and lump all heuristics under discovery. Note the stark contrast between this and the experiments discussed in the first six chapters of this book, which focus more on the role of such heuristics as disconfirmation in hypothesis testing. Furthermore, whereas the early/late distinction was discovered in laboratory simulations, Langley et al. propose to determine what constitutes good discovery heuristics by simulating historical "breakthroughs" on the computer. Their goal is "to explore the role in scientific discovery of heuristics that may be relevant over a wide range of scientific disciplines and hence may contribute to our basic understanding of discovery wherever it may occur. In adopting this strategy, it is not our intention to deny the important role that discipline-specific knowledge and heuristics play in the work of science; rather, we want to see how far we can go initially with data-driven, semantically-impoverished processes" (Langley et al. 1987:65). Their focus on general, "semantically-impoverished" heuristics is similar to the focus of the original 2-4-6 and Eleusis experiments; like BACON, these tasks do not incorporate the extensive declarative and procedural knowledge of the working scientist. These programs, like the experiments, can be used to make externally valid claims.

Langley et al. apply these heuristics to emulate the sorts of discoveries that occur in semantically rich environments, instead of problems like the 2-4-6 task that simulate features of scientific reasoning. In other words, they want to use more ecologically valid data. But in taking this apparently beneficial step, they make a mistake: they confuse the cleaned-up data presented to the program with the "data" discovered or created by the scientist at the time. They then make a claim that blurs the external/ecological distinction: that these programs actually discover scientific laws. Do they mean that BACON shows how this set of heuristics might discover scientific laws under ideal circumstances, using "cleaned-up" data (externally valid claim)? Or do they imply that BACON models the actual process by which Kepler and others found their scientific laws (ecologically valid claim)? Computational simulations can have enormous heuristic value, provided one is clear about one's goals.

To illustrate these issues, let us consider BACON.1, the most primitive version of BACON. One of this program's goals is to simulate data-driven discovery by applying heuristics to two columns of numbers representing the distance of each planet from the sun and its period of revolution. Another goal is to do what Kepler did. These two goals are not the same.

BACON.1 is given data on distance and orbital period for each planet; it does not have to discover these values, as Brahe, Kepler, and others did. Then it applies three heuristics to the columns of data: (1) "if the values of a term are constant, then infer the term always has that value"; (2) "if the values of two numerical terms increase together, then consider their ratio"; (3) "if the values of one term increase as those of another decrease, then consider their product" (Langley et al. 1987:66–70).

Figure 9-3 illustrates this process. Note that the three question-boxes, representing BACON.1's heuristics, are not connected in linear fashion as in a top-down flowchart. BACON.1's heuristics are represented as produc-

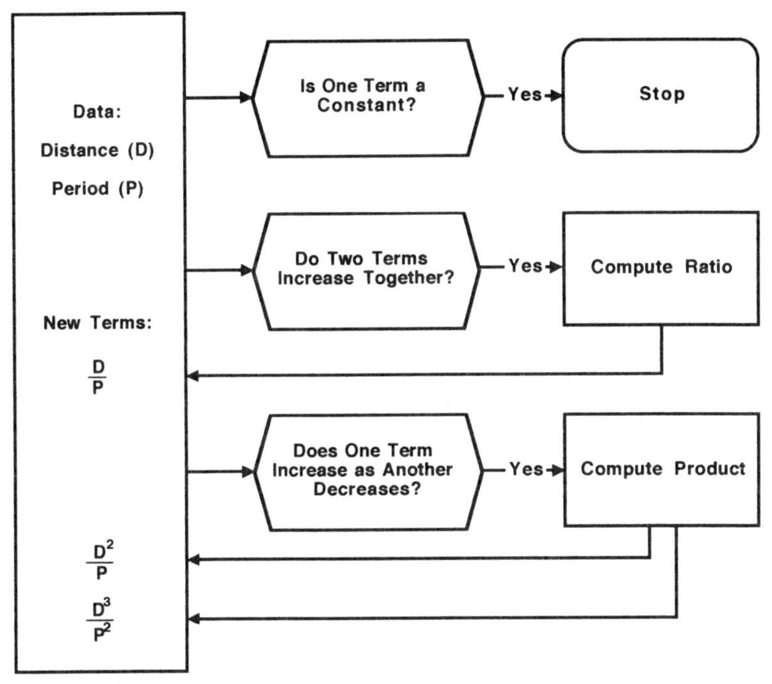

FIGURE 9-3 BACON.1'S HEURISTICS

tion rules, which "fire" whenever an inspection of the data satisfies their conditions. The distance (D) and period (P) increase linearly, so the second heuristic fires and creates a new term, D/P, which is added to the data set (see Figure 9-4). Now the D column increases as the D/P column decreases, so the third heuristic creates a new term, D^2/P. D/P and D^2/P vary inversely, so the same heuristic fires again, creating D^3/P^2. This new term has a constant value of 1 for every planet, which causes the first heuristic to fire and stops the iterative process.

According to Langley, Simon, Bradshaw and Zytkow, BACON.1 has now independently discovered Kepler's third law. Indeed, Simon (1988) wondered why it took scientists like Kepler so long to make these discoveries, when a computer could do them in a few minutes, at most.

Planet	D	P	$\frac{D}{P}$	$\frac{D^2}{P}$	$\frac{D^3}{P^2}$
A	1.0	1.0	1.0	1.0	1.0
B	4.0	8.0	0.5	2.0	1.0
C	9.0	27.0	0.333	3.0	1.0

FIGURE 9-4 DATA FROM THREE HYPOTHETICAL PLANETS (Based on Langley, Simon, Bradshaw, and Zytkow 1987:69.)

9.2.1. Does BACON Discover?

When Langley et al. say that BACON discovers, are they making an externally or ecologically valid claim? The authors conclude that their programs demonstrate "ways in which historical discoveries could have been made with the help of a few simple heuristics and with moderate amounts of computation" (1987:170). This sounds like an externally valid claim: BACON.1 shows that three heuristics are sufficient to reproduce the mathematical relationship that underlies Kepler's laws.

But one cannot make the ecologically valid claim that the program reproduces Kepler's path to discovery. Kepler was not given two columns of numbers and a set of heuristics that pointed him toward the right relationship. Indeed, he had to decide whether to take Brahe's data seriously and struggle to find a numerical relationship, beginning with the idea that perfect solids should fit inside the planetary orbits. Along the way, he had to abandon the Copernican dogma that all the planetary orbits had to be circular (see Koestler 1963 and Bernstein 1978). He had to replace the old order with a new one, based on the idea of an "*anima motrix*" emanating from the sun and on William Gilbert's magnetism (see Kuhn 1957). That Kepler's successors took only his mathematical description of the planetary orbits seriously does not mean that his discovery consisted solely of a clever set of numerical manipulations.

However, can one make the normative argument that Kepler would have discovered his laws faster had he used these three heuristics? Here the translation of externally valid evidence into a normative claim is tricky, and depends on further research.

One possibility would be to train subjects to use BACON.1's heuristics and see if they solve a variety of problems faster than control subjects can. Qin and Simon (1990) asked fourteen college students to work on columns of data similar to those given to BACON.1; only four succeeded in finding the mathematical relationship we now call Kepler's third law, and they tended to use more of the heuristics employed by BACON than did other subjects, though there were other reasons for success and failure, e.g., some subjects lacked appropriate mathematical skills. The authors conclude that

> Kepler's discovery of his third law was an event of great significance in the history of science. It is regarded as a discovery of the first magnitude. From the fact that, with given data, 4 out of 14 subjects could rediscover this law within one hour, and from the search processes revealed in the subjects' protocols, it can be said that significant discoveries can be made simply by application of the general processes that have been observed in all kinds of problem solving. (Qin and Simon 1990:303)

Here is the clearest statement of the normative claim. Had poor Kepler only known about these heuristics, he would have discovered his third law in a trice! Later in this chapter we will discuss what kind of additional experimental evidence is needed to check this claim. But for now, let us make the obvious point that we cannot know what could have been done to "speed up" Kepler's process. The closest we could come would be to assess the value of such heuristics for modern scientists.

Qin and Simon next try to derive an ecologically valid claim from their research, speculating about "what light such an experiment could cast on an actual historical instance of discovery in a case where there is reasonably good evidence that the discovery was driven by the data and received no substantial guidance from relevant preexisting theory" (1990:305). Now they are making the assumption that Kepler's mental representations played almost no role in his actual discovery. But had Kepler not been a Copernican, he never would have searched for his laws. It is true that he modified his theories in significant ways to accommodate the data, but he did not merely look at two columns of numbers and explore an atheoretical relationship. Indeed, he had an evolving mental model of the solar system that guided his efforts. Later, Qin and Simon admit that "not too much detail is known about how he derived the third law" (p. 305) and that "he did not leave behind a record of the heuristics he used" (p. 307). So their ecologically valid claim rests on insubstantial evidence.

Qin and Simon's experiment and Langley et al.'s programs certainly provide externally valid evidence that BACON's heuristics are useful in discovering numerical relationships. But the authors only confuse matters by claiming that BACON discovers. As Langley et al. admit, many aspects of discovery are not simulated by BACON.

Figure 9-5 illustrates a number of activities that might be grouped under discovery. BACON.1 can simulate heuristics that might be useful in manipulating numbers to discover relationships between columns of data. But unlike Kepler, Einstein, Newton, Darwin, and other scientists, BACON.1 doesn't have to formulate a novel problem; its problems are given to it. Furthermore, the program has no mental image or model of the solar system which can guide or inhibit its discoveries: it ignores the whole heliocentric controversy. Finally, it is the programmers who decide that BACON.1 has discovered Kepler's laws: neither the program nor Qin and Simon's subjects know the significance of the relationship they have discovered.

Later versions of BACON include additional heuristics designed to make it applicable to other scientific discoveries. For example, BACON.4 has the ability to use nominal variables, i.e., ones that have labels instead of numbers as their values. The variable "material," for example, might have lead, silver, and iron as nominal values. BACON.4 could derive the intrinsic property "density," given appropriate data, then apply heuristics similar to those of BACON.1 to discover laws regarding these intrinsic properties. For example, BACON.4 is able to derive an equation similar to Ohm's law given data for nine combinations of batteries and wires; with battery

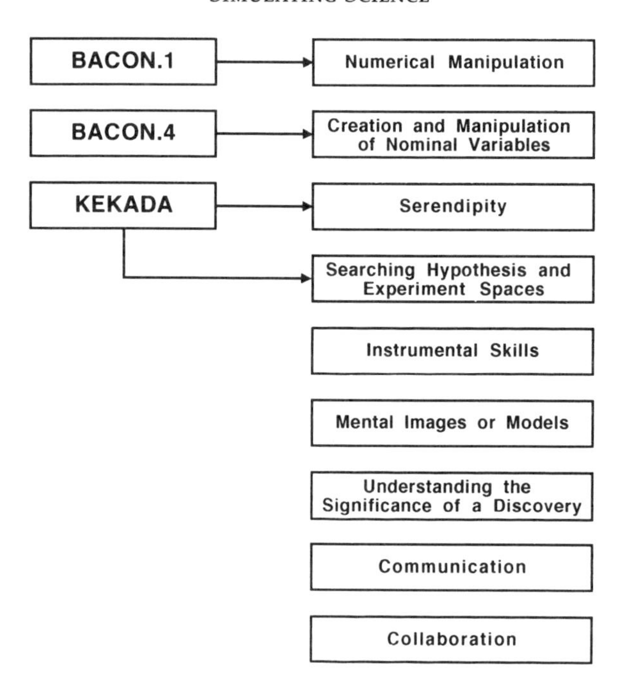

FIGURE 9-5 ASPECTS OF DISCOVERY

as one nominal variable and type of wire as another, it discovers intrinsic properties: conductance associated with batteries and voltage associated with wires.

While these refinements increase the generality of the externally valid claims that can be made from these programs, allowing them to include more aspects of discovery (see Figure 9-5), there is no concomitant increase in ecological validity. As the authors admit, the program does not know what these intrinsic properties mean. It has no mental representation of batteries, wires, or electricity (see the discussion of Gentner's research on how mental models affect solution of electrical problems in 10.3.4). The computer simply runs a set of production rules until it reaches a state that causes a rule to fire that stops further calculation. It is the programmer who recognizes that a discovery is made and who explains that the discovery is about orbits, or batteries and wires.

9.2.2. WILL THE NEXT KREBS BE A COMPUTER?

Kulkarni and Simon (1988) designed a program called KEKADA to simulate an ecologically valid situation: Krebs's discovery of the ornithine cycle, which was due, in part, to his detecting and pursuing the cause of an anomalous result. Krebs noticed "the striking rise in the rate at which a liver slice produced urea when the specific amino acid ornithine, together with ammonia, was added to its medium" (Holmes 1987:228).

Therefore, KEKADA incorporated one new aspect of discovery: the ability to take advantage of this sort of serendipity (see Figure 9-5).

To accomplish this, the program divides the exploration process into separate searches through hypotheses and data. KEKADA is given the result of an experiment and expectations for that result; when a contradiction or "surprise" occurs, the program generates new hypotheses and experiments.

The idea of a dual-space search was derived from experimental research. For example, Klahr and Dunbar (1988) asked subjects to guess the function of the RPT key on a device called a Big Trak. They found that subjects adopted one of two cognitive styles: theorist or experimenter. Theorists changed their representation of the RPT key's function by searching the hypothesis space; experimentalists switched representations by exploring the experiment space until they encountered a disconfirmatory result. Klahr and Dunbar used this evidence to argue for the external validity of a distinction between hypothesis and experiment spaces.

Kulkarni and Simon used this distinction to divide heuristics into two types: one set searched the hypothesis space to determine which hypothesis KEKADA should focus on next, and another set searched the experimental space to come up with experiments that would test the current hypothesis. Results of experiments "are supplied interactively by the user" (Kulkarni and Simon 1988:156). Kulkarni and Simon made sure that KEKADA was particularly attuned to following those experiments or results that disconfirmed expectations. Decisions about what hypothesis to pursue at certain points in the program were also supplied by the user.

The authors made another distinction between types of heuristics, arranging them in a hierarchy from general to specific (see 3.6.2). They "were able to show that nearly half of the heuristics Krebs used were quite general, being relevant not only beyond the Urea synthesis problem, but beyond chemistry to a wide range of research situations" (1988:173); the other half were domain-specific. The boundary between general and specific was left somewhat fuzzy by the authors, but the implication is that general heuristics could be used across several scientific disciplines, whereas domain-specific were limited to a single discipline or area within a discipline.

Consider one of KEKADA's general experiment-proposing heuristics: "If the goal is to study a particular reaction in general, carry out the reaction under various conditions." A similar domain-specific heuristic is, "If you are studying a phenomenon with A as reactant, and there is a hypothesis that A and B react to form C, carry out experiments on A and on B, and compare rates of formation of C from A and from B" (p. 155).

There is a third level of heuristic: those specific to the individual scientist. Kulkarni and Simon "found that Krebs' choice of problem and technique were much determined by the special opportunities provided by his training in Otto Warburg's laboratory. The tissue culture method, acquired there, was his 'secret weapon', his source of comparative advantage" (p. 173). This third level illustrates one of the problems with KEKADA: there is no way in which it can incorporate the instrumental skills which Kulkarni and Simon identify as the critical element in Krebs's success. In general, programs cannot duplicate the kind of hands-on, craft knowledge possessed by scientists; they cannot manipulate instruments, identify substances by sight or smell, or refine delicate experimental procedures. KEKADA, therefore, cannot simulate Krebs's "secret weapon."

Furthermore, it cannot simulate the way in which the act of trying to communicate his discoveries improved Krebs's understanding of the ornithine cycle. Holmes (1987) argues that Krebs may have written about two-thirds of his first paper on urea synthesis before he discovered the ornithine effect. Krebs himself said, "I spent a lot of time on writing, but usually while the work was still going on. And I find in general only when one tries to write it up, then do I find the gaps. I cannot complete a piece of work and then sit down and write the paper" (Holmes 1987:226).

KEKADA certainly covers more aspects of discovery than BACON (see Figure 9–5). Unlike BACON, it is supplied with a larger base of declarative knowledge specific to chemistry; therefore, it can incorporate domain-specific heuristics as well as general ones, and can trace a discovery path that comes closer to a historical case than can any of the versions of BACON. But like BACON, KEKADA is supplied much of the knowledge and constraints it needs either in advance or interactively by the user; it is led carefully along Krebs's path. KEKADA literally does not know what it is doing; it does not realize what it is discovering.

Once again, the issue is one of claims. Kulkarni and Simon want to use KEKADA to claim (1) that scientific discovery is a gradual process, not just a flash of genius; (2) that scientific discovery depends on several levels of heuristics, like other forms of problem solving (see Kulkarni and Simon 1988:174). These externally valid claims are supported by their program as well as by other research on heuristics (cf. Perkins 1981 for the first claim and much of the work in earlier chapters for the second).

Kulkarni and Simon also make the ecologically valid claim that their heuristics, in combination with sophisticated declarative and procedural representations and instrumental skills, might have played a major role in the discovery of the ornithine cycle. But KEKADA does not simulate the interaction between heuristics, skills, and these other forms of representation. Even if it did, to evaluate this claim we would have to look at Krebs's processes in very fine detail.

9.2.3. EXPERIMENTAL AND COMPUTATIONAL SIMULATIONS COMPARED

Klahr, Dunbar, and Fay (1990) comment that while computational simulations of actual discoveries appear to have greater "face" validity than experimental studies using students and abstract tasks have, the latter permit more precise control of the context of discovery and finer-grained observations of cognitive processes. "Face" validity corresponds to what we are calling ecological validity, and the issue of control relates to external validity. The computational simulations appear to be more ecologically valid because they are given data similar to that used by actual scientists who made important discoveries. But as we have seen, this ecological validity is in fact quite limited.

Computational simulations potentially have even greater external validity problems than experiments have. In a sense, Langley et al. merely substitute a machine for a human subject. The advantage is that they can control and manipulate the machine's "thought processes"; the disadvantage is what they want to generalize from the machine to the human case, and this raises additional problems. There is no guarantee that heuristics which are successful on a computer will work for human beings, even if the heuristics are derived post hoc from a case of discovery. Experiments can also explore what kind of representations human beings form as they try to apply these heuristics. That is why studies like the one by Qin and Simon (1990) are essential complements to such programs as KEKADA and BACON.

But the Qin and Simon study raises another issue. Are computational simulations more likely than experiments to produce confirmations? To put it another way, are programs less likely than experiments to produce the kind of surprises that led to the discovery of the ornithine cycle? Experiments often "fail" because of the recalcitrance of human subjects, who refuse to respond in accordance with prevailing theory. Every programmer knows that computers are also recalcitrant and that programs can surprise their creators, but the Qin and Simon study suggests that there is a difference between experiment and program. Experimental results generally confirmed that heuristics such as BACON.1's can be used to discover patterns in columns of numbers. But subjects used a wider range of heuristics and not all provided confirmatory evidence; Qin and Simon note that at least one subject who applied the correct heuristics failed to find the relationship, because he searched for the wrong type of function. One could argue that this subject represented the problem differently, and that his or her result illustrates the limits of BACON as a model of human problem-solving.

This observation suggests an additional line of research that could explore the external validity of the claims made from BACON. Suppose we deliberately instructed some subjects to use BACON's heuristics, and compared them to a group instructed to represent the data as a search for a relationship between distances and orbits. (Care would have to be taken to ensure the subjects did not know Kepler's law.) Judging from Gorman, Stafford, and Gorman's (1987) results, those subjects in the representation condition might perform as well as, or better than, those in the heuristic condition. The representation subjects might even independently evolve some of the heuristics used by Qin and Simon's subjects, but apply them with more efficiency and success. Such a result would support our earlier conclusion that successful application of heuristics depends on achieving an appropriate task representation (see sections 3.4 through 3.6).

How could such a line of research be linked with experiments using tasks such as Eleusis and the 2-4-6? One could transform Qin and Simon's problem into a task more like the 2-4-6 by giving students one or two examples of the relationship between D, P, and D^3/P^2. Subjects could then be asked to propose additional triples in an effort to discover the rule, using a calculator to explore mathematical relationships. Again, one could compare a variety of instructions designed to affect the way subjects represented the problem and/or what heuristics they employed, including the extent to which they relied on confirmation and disconfirmation.

To make this problem more like BACON's task, one could give subjects examples of D and P and ask them to generate triples that included a third term, which would correspond to D^3/P^2. One could add possible or actual system-failure and/or measurement error to explore its effects on subjects' abilities to apply BACON's heuristics. One could also study how subjects represent the rules they are seeking, and even compare the effect of manipulations designed to alter those representations.

Finally, one could employ groups in a design of this sort, investigating the effect of such variables as error and minority influence on consensus and knowledge transmission. Communication and collaboration or debate among scientists are important aspects of science that BACON and KEKADA cannot simulate (see Figure 9-5). As noted earlier, Holmes (1987) shows how both Lavoisier and Krebs made new discoveries in the course of writing about their research. Modern discoveries are often made by research teams (cf. Holton 1973); BACON itself was developed by such a team. Even Kepler's discoveries were influenced by interactions with colleagues such as Brahe and Maestlin.

The ability to simulate group processes is one area in which experiments are definitely superior to computational simulations, although, as we shall see below, at least one "computationalist" contends that computers could simulate social interaction. Even some of the simpler aspects of written communication could be simulated in the laboratory, including aspects of the referee system used by most journals (see 7.3.3). But experimental methods need to be supplemented by historical and modern studies of working scientists, including rhetorical analyses.

Consider, for example, the rhetoric used by Langley, Simon, and others associated with these computational simulations. Why do they claim these programs discover, when it is clear that they simulate only a portion of the activities often listed under discovery (see Figure 9-5)? The authors know most of the limitations of their programs and discuss them at length. But they also know that claiming their programs simulate heuristics that might be useful in scientific discovery will attract less attention than claiming they have developed programs which discover. The latter makes it sound like BACON and KEKADA have at last fulfilled some of the extraordinary predictions about Artificial Intelligence made by Simon and Newell in 1958.

9.2.4. Correcting and Augmenting Scientific Theories

In contrast, other authors of computational simulations designed to model aspects of scientific reasoning are more careful about their claims. For example, Rajamoney (1990) designed a program called COAST that takes over where discovery programs like BACON and KEKADA leave off. COAST takes a newly discovered theory, still rudimentary and ill-formed, and revises it; the program detects anomalies, revises the theory to account for them, and designs experiments to test the revisions. COAST represents theories as collections of exemplars; any revision must still account for the exemplars that were explained by the earlier version of the theory. Rajamoney is careful to distinguish the sorts of revisions COAST makes from the revolutionary conceptual changes that occur in times of Kuhnian crisis. COAST's cases or problems are not typically the historical cases used by BACON and KEKADA, though Rajamoney plans to extend it to such problems; instead, it is designed to deal with a wide variety of scenarios, though Rajamoney's examples are all fluid problems. For example, he shows that COAST can develop a new process definition that approximates osmosis. According to Rajamoney, the major difference between KEKADA and COAST is that the former uncovers new anomalies and explores them while the latter uses anomalies to spur theory revision.

Rajamoney is clear about COAST's limitations: it forces theories to assimilate observations, and therefore cannot simulate a situation in which a theory is discarded and replaced by a rival. COAST has no way of representing competing theories. But COAST does present us with an interesting model of how theory revisions might occur under conditions that Rajamoney needs to specify more precisely.

The program needs to be complemented by evidence from experimental and historical cases. It would be interesting, for example, to run COAST on some of the situations encountered by subjects on Eleusis or on the 2-4-6 task. One could give COAST a subject's representation of the problem after she or he had formulated a tentative hypothesis, and present the program with some of the anomalies actually encountered by the subject. The dialogue between program and experiment might lead to a more sophisticated model of hypothesis revision. On the historical end, one might explore the ecological validity of COAST's methods by comparing them with detailed analyses of actual cases of theory revision.

9.2.5. Increasing Ecological Validity

Klahr, Dunbar, and Fay (1990) argue that there ought to be more observation of "fine-grained" cognitive processes. The work of Tweney and Gooding on Faraday (see section 8.1) shows how this might be done.

Gooding (1990), in particular, has constructed detailed maps or charts of Faraday's experimental activities that resemble Newell and Simon's problem-behavior graphs.

It may be possible to construct a future FARADAY program based on research like that of Gooding and Tweney, though the "hands-on," procedural components of knowledge will be particularly hard to model. As it stands, Kulkarni and Simon indicate that they are testing their model of discovery on other problems, including Krebs's discovery of glutamine synthesis. But their efforts thus far have fallen far short of the level of fine-grained detail reached by Gooding. One would hope they would improve their simulations either by doing original historical research or by collaborating closely with a historian. This sort of work might get around the problem outlined by Suchman (1987), who claims that cognitive scientists often make the mistake of assuming that actions are governed by heuristics because these actions can be described after-the-fact as having followed heuristics (see also Collins 1990). Working with historians and/or original historical data would help cognitive scientists avoid this sort of oversimplification, in part because they would gain a more complete picture of how the scientist or inventor represented her or his problem at the time. Chapter 10 will outline a program of research into the fine-grained cognitive processes of inventors and will include suggestions on how the computer might be used to assist the historian.

9.3. WILL THE NEXT KUHN BE A COMPUTER?

Recently, philosophers have picked up on Simon's idea of using computer simulations to develop better norms for scientific practice. Darden (1990), for example, is working on developing a kind of expert system that might assist scientists in diagnosing faults in theories. Karp (1990) has constructed a program called HYPGENE that designs scientific hypotheses using an extensive knowledge base; it could potentially be adapted to play an expert assistant's role. Both Darden and Karp simulate situations derived from genetics, but both clearly hope their programs could eventually work in a variety of scientific domains. Instead of replacing Kepler, such programs might possibly have helped him develop alternate hypotheses for the anomalous behavior of Mars.

Paul Thagard has an even more ambitious goal. He wants to use computational methods to settle the question of which philosophy of science works best:

> If competing methodologies were developed explicitly enough to be programmed, we could compare a group of conservative Kuhnian scientists with a group of more critical Popperian scientists; my conjecture is that in the long run the Kuhnian group would accomplish more. Even more interesting, we could consider a mixed group with different methodological styles, encom-

passing the audacious, the critical, and the conservative. The experimental results might be an important contribution of computational philosophy of science to the theory of group rationality in science. At the very least, the exercise of working out the nature of such methodologies in sufficient detail to be implemented in a computational model would be highly illuminating. (Thagard 1988:188)

Thagard's language suggests that he wants to use computational simulations to make externally valid claims about the relative merits of philosophical approaches to science. This is a perfectly legitimate use of simulations—provided other forms of evidence are used to check the normative value of the claims. The experimental approaches discussed so far in this book raise problems with both naive and methodological falsification (see chapters 3 and 6), but they do not "refute" them. Instead, they demonstrate that the effectiveness of such heuristics as confirmation and disconfirmation depends on relationships between representation, target rule, and level of error.

In other words, the cognitive approaches discussed so far do not settle the issue of whether Popper, Lakatos, or Kuhn provides a better overall model of the growth of science, partly because these philosophical frameworks are all extremely general and cognitive methods have typically been used to elucidate what happens under specific conditions. For example, in specific situations one can recommend such heuristics as disconfirmation and replication-plus-extension, but these are not universal norms.

9.3.1. Process of Induction: A Connectionist Alternative

To accomplish his goal, Thagard relies on a different type of computational simulation. One of the major weaknesses of BACON and KEKADA is that, like the early 2-4-6 and Eleusis studies, they place too much emphasis on heuristics and too little on knowledge transformation. KEKADA's declarative knowledge grows in the course of its simulation, but it still has no visual representation or mental model of what it is doing (see the next chapter for a more detailed discussion of mental models).

"Process of Induction" (PI), a program developed by Holland, Holyoak, Nisbett, and Thagard (1986), is a rule-based system that tries to take this mental model issue into account. As the authors note, in a critique of BACON, "Although he started with relatively inaccurate data furnished by Tycho Brahe, Kepler's discovery of the three famous laws would have been impossible without several dramatic reconceptualizations of the problem domain" (p. 324). In other words, Holland et al.'s view of scientific problem-solving comes closer to the view articulated in chapter 3, which discussed the role of mental representations.

There are many types of connectionist systems, but all involve networks of interlinked units; knowledge is embodied in patterns of connections, not in the individual units, any one of which may be involved in multiple pieces of knowledge. In these systems, concepts or rules emerge as statistical properties of patterns of activation (see Holland et al. 1986). These networks can learn by being presented with examples; gradually, the system will settle on patterns of activation consistent with its experience but flexible enough to incorporate new evidence. Processing of information occurs in parallel; more than one rule or concept can be active at a given time. The idea of units and connections is a rough parallel with the human nervous system; in fact, a particular kind of connectionist system is referred to as a "neural net" and at least one philosopher sees these neural nets as the solution to major problems in the philosophy of mind (Churchland 1989). Connectionist networks are particularly good at simulating the reconstructive aspects of memory (Johnson-Laird 1988) and categorical representations that center around prototypical instances rather than discrete lists of attributes (Holland et al. 1986).

How might Holland et al. model Kepler's problem-solving process? Their implementation of PI was very limited and could not handle anything like the Kepler and Krebs cases, but one can speculate, given their description of their program. Presumably, a PI simulation called KEPLER would have to begin with a set of mental models, represented by interconnecting rules. Unlike newer, more sophisticated connectionist systems (see Waltz 1988 for an example), it is not clear that PI could learn these rules "from scratch"; instead, the system seems more oriented to modifying and creating new rules out of existing ones. Perhaps we should start KEPLER with one of Kepler's early mental models: that planetary orbits should fit inside a progression of geometric solids with increasing numbers of sides, and include subordinate rules for each planetary orbit. As Brahe's data were entered, the weights associated with each subordinate rule would change; the one for Mars would present the worst fit, and would have to be adjusted or abandoned. Eventually, the whole system of geometric solids would be replaced by a competing model. Perhaps PI could arrive at the idea of an oval orbit on its own, or perhaps that would have to be provided by the programmer as an alternative. It is not clear. However, it is easy to imagine an ellipse model outcompeting a perfect-circle rival.

The Kepler example illustrates some of the potential strengths and weaknesses of PI vis-à-vis BACON and KEKADA. Like BACON and KEKADA, PI would have to rely heavily on the programmer to set up initial conditions and provide background knowledge, though in PI's case, it is possible that Kepler's initial assumptions could be represented as sets of rules and the system could take the data in bits and pieces, the way Brahe fed it to Kepler, and modify its representations as it went along.[20]

Like BACON and KEKADA, PI relies solely on propositional representations, whereas Kepler saw and imagined hearing the harmony of the spheres. As Kosslyn (1980), Anderson (1983), Paivio (1986), and Finke

(1989) have shown, visual representations are distinct from propositional representations and play a crucial role in human cognition (but see Chambers and Reisberg 1985 for a counter-argument); as Giere (1988), Miller (1989), and others have demonstrated, visualization is critically important in science.

So, it seems that neither BACON nor KEKADA nor PI will become the next Kepler or Krebs. Perhaps one of the new connectionist or neural-net computers, currently in their development phase, could more closely approximate scientific discovery. These machines would be capable of acquiring knowledge on their own, eliminating the need for the programmer to supply the expert's knowledge.

Like the experiments, however, the relatively simple programs outlined above can help formulate externally valid generalizations about processes that are important in scientific discovery, for example, by suggesting heuristics that might facilitate discovery and/or by suggesting how theories can be viewed as mental models with subordinate rules. Like the 2-4-6 and Eleusis experiments, results from these computational simulations may be relevant to issues in philosophy of science.

9.3.2. ECHOing Lavoisier

Thagard (1989) developed a program called ECHO, based on PI, to simulate the oxygen-phlogiston controversy. His goal was to confirm his philosophical position that in the competition between theories, ones with superior explanatory coherence win. Though Thagard states explanatory coherence in terms of a complex network of propositions, the basic idea is simple: the most coherent theory explains the most data with the fewest corollary hypotheses and internal contradictions.

ECHO is given rules in the form of LISP code; (EXPLAIN (H1 H2) E1) means that Hypotheses One and Two explain a piece of Evidence arbitrarily labeled One. Note that the program has no idea what these hypotheses or data represent; like BACON, it does not know what it is doing. But given a set of relations of this sort between arbitrary Hs and Es, it can establish a network of Hs that connect and explain the Es, which will involve eliminating some of the Hs and even some of the Es.

Then Thagard sets up the oxygen/phlogiston controversy in terms of Hs and Es, based on his reading of Lavoisier's 1783 polemic against phlogiston. Thagard admits that one could reconstruct the whole controversy from the standpoint of a phlogiston theorist, suggesting that one could input the Hs and Es so that a phlogiston account would emerge as more coherent than an oxygen one. Thagard justifies his bias toward Lavoisier by pointing out that most chemists and physicists eventually accepted his claims. But ECHO tells us nothing about why they accepted these claims; it merely shows that an abstract version of Lavoisier's own argument, programmed on a computer, does in fact support his theory.

Thagard uses ECHO in a similar way to show that Darwin's justification of evolutionary theory has more explanatory coherence than rival accounts have. Recently, Thagard has used ECHO to show why Wegener's theory of continental drift was rejected in the 1920s and a more sophisticated version of it accepted in the late 1960s (Thagard and Nowak 1990). More intriguingly, Thagard uses ECHO to predict the resolution of a scientific controversy, assessing several rival hypotheses concerning the extinction of the dinosaurs, based on conversations with one paleontologist. The program does not eliminate any of the rivals, but it gives the greatest explanatory coherence to the Alvarez hypothesis. What isn't clear is whether the program is merely reproducing the perspective of this one scientist.

9.3.3. ECHO AND THE RHETORIC OF SCIENCE

Giere (1989b) makes the point that what ECHO really models is the structure of Lavoisier's and Darwin's successful arguments. In other words, ECHO provides a computational representation of Lavoisier's persuasive rhetoric and shows that if you formulate the hypotheses and evidence in just the way that Lavoisier requires, his theory "wins." As Thagard himself points out, "We would get a different network if ECHO were used to model critics of Lavoisier . . ." (1989:446). So one is left wondering just what it is that ECHO demonstrates—that both Lavoisier and his critics can develop persuasive arguments that can be represented in a computational network?

Thagard appears to want to model both the way scientists ought to reason (external validity) and the way they actually reason (ecological validity). For example, he argues that "we are trying to model the actual thought processes with which great scientists initiated scientific revolutions" (Thagard and Nowak 1990:28). But while ECHO may incorporate Lavoisier's persuasive rhetoric, it in no way models the "thought processes which lead him to initiate a revolution." Holmes (1987) documents some of the twists and turns. Lavoisier's "Experiments on the Respiration of Animals and on the Changes Which Take Place in the Air in Passing Through Their Lungs" was presented to the Academy of Sciences in 1777 and was subsequently published. The finished paper presents experiments on the calcination of mercury which indicate that "dephlogisticated air" (oxygen) combines with mercury to create "a red calx," and also experiments on respiration in birds which indicate that "dephlogisticated air" is converted in the lungs to an equal amount of "fixed air." Close inspection of one of his early drafts reveals that he initially saw a compelling analogy between "red calx" and red blood:

> One cannot doubt, therefore, that the respirable portion of the air becomes fixed in the lungs, and it is very probable that it combines there with the blood. That probability is transformed to a kind of certitude if one considers that it is a recognized propery of the air that is better than common air to impart a red color to the bodies with which it combines. Mercury, lead, and iron provide examples. (quoted in Holmes 1987:224)[21]

In a later draft, Lavoisier proposed two alternatives: "one, either a portion of . . . dephlogisticated air . . . is absorbed during respiration and replaced by an almost equal quantity of fixed air or mephitic gas which the lungs restore to its place, or else the effect of the respiration is to change that same portion of dephlogisticated air into fixed air in the lungs" (quoted in Holmes 1987:224). Eventually, after more redrafting, he settled on the second alternative.

As Holmes points out, we cannot tell whether these conceptual changes occurred in the act of writing or in moments of reflection between writing sessions, but "the timing is, I believe, less significant than the fact that the new developments were consequences of the effort to express ideas and marshall supporting information on paper" (Holmes 1987:225). Significant knowledge transformations occur during the act of writing, and neither experimental nor computational simulations capture this feature of science.

Perhaps the appearance of explanatory coherence is a useful rhetorical device for persuading others. One could then present Thagard's principles of explanatory coherence (1989:436–438) as heuristics for use in writing persuasive scientific papers, and explore whether journals tend to publish scientists who use them.

9.3.4. BACON, KEKADA, AND ECHO

Simon (1989:487) compared Thagard's ECHO simulation of the oxygen-phlogiston controversy to one of BACON's "cousins," STAHL, a program which also was designed to simulate the oxygen-phlogiston controversy (see Langley et al. 1987) Simon argued that STAHL worked with many fewer "givens" than ECHO: whereas ECHO has to be given both hypotheses and evidence, STAHL is given only heuristics and evidence, and has to derive the hypotheses. In both cases, this evidence has been "cleaned up," which limits ecological validity.

Simon also pointed out that STAHL is more domain-specific than ECHO is, because the former's heuristics are tailored toward chemistry whereas the latter "is quite general, but only because all of the domain-specific knowledge is provided to it by the user in each application and it is oblivious to the content of its propositions" (1989:487). One could argue that STAHL, like BACON, is "oblivious to the content of its propositions" as well, though it is supplied with variable names that denote substances like "calx-of-iron" and "phlogiston ash": as far as the program is concerned, these "substances" are merely arbitrary variables. Both programs also ignore the important hands-on skills possessed by experimental scientists and assume that the propositional aspects of science can be modeled independently of the visual and tactual dimensions.

Simon's final point is that "Both STAHL and ECHO will corroborate the oxygen theory of combustion if given Lavoisier's 'facts' and the phlogiston theory if given STAHL's 'facts' " (1989:487). To illustrate his point, let us consider another example: programming ECHO to simulate the canals-on-Mars controversy from Lowell's perspective (see 3.2). We could arrange the hypotheses and evidence presented to the program so that it would reproduce Lowell's perspective. From his standpoint, the canal hypothesis had greater explanatory coherence than its rivals. After all, he tied together in a single hypothesis a group of diverse "facts": the canals waxed and waned in inverse relation to the Martian polar caps, they were consistent with the presence of a small amount of water vapor in the planet's atmosphere, and new canals appeared in a way that supported Lowell's theory. Opponents attacked Lowell's "facts," dismissing his observations as optical illusions, but he even had a theory to account for why his observatory was better suited to Martian viewing than were those of his rivals. Besides, critics had to explain why Lowell was not the only one to see these canals. It should by now be obvious that one could program a computer using the evidence as seen from a Lowellian perspective and it would arrive at the conclusion that his theory was more coherent than that of his rivals.

One might even be able to set up an ECHO simulation that showed that Von Daniken's theory about extraterrestrial influences on human prehistory (see Introduction) had more explanatory coherence than its rivals. After all, from Von Daniken's perspective, his theory explained all sorts of anomalous data. Perhaps explanatory coherence is really a theory of human rationalization, not reasoning.

From BACON, KEKADA, and STAHL, we can at least derive externally valid heuristics that could be used, depending on the data, to derive either oxygen or phlogiston theory. Similarly, Thagard hopes that ECHO leads to the externally valid conclusion that explanatory coherence will distinguish good theories from bad ones. But the Lowell example above suggests that explanatory coherence may be in the mind of the beholder.

9.3.5. The Emperor's New Epistemology?

All of this computational simulation demonstrates that (surprise!) Thagard's principles of explanatory coherence provide rational criteria for hypothesis evaluation. Indeed, Thagard uses ECHO in a way that reminds one of Lakatos's (1978) rational reconstructions (see 3.9). Thagard believes historiography must be influenced by the methodological conclusions one intends to draw; Lakatos believes history must be read in the light of what we know about rationality. Both intend to make normative claims, based on their reconstructions. The difference is that Thagard makes it appear that the rational reconstruction simply emerges from an objective machine. In both cases, we run the risk that history will be distorted to fit philosophical or methodological preconceptions. The danger is that Thagard may have invented the perfect confirmation machine: input your philosophy, and watch it predict the winner.

To put it in experimental terms, Thagard has used no controls; he has not demonstrated that an alternative philosophy would not provide an equally good account of the oxygen/phlogiston controversy. Perhaps we need programs called KUHN, LAKATOS, and POPPER, programmed by adherents of each of these philosophical views, to compete with Thagard's.[22] Presumably, a Kuhnian simulation would support an incommensurability account of the controversy, a Lakatosian simulation would demonstrate that phlogiston was a degenerating program and oxygen a progressive one, and so on.

There is, of course, a similar danger with experiments; as McGuire (1989) noted, a clever experimenter can arrange results to come out in accordance with expectations (see 7.3.1). Fortunately, human subjects (and animals) often surprise the experimenter, creating the sorts of paradoxes that fueled my research program. But clearly, confirmation bias can be a problem for both experimental and computational simulations. One solution, as we have noted earlier in the book, is to design disconfirmatory simulations. For example, from Thagard's research one would predict that a computational simulation of the oxygen/phlogiston controversy based on some other approach besides explanatory coherence should fail. That is easily tested, with the same or another program—and in fact, STAHL already provides a potentially negative result.

But we are still left with a larger question. Why substitute computer programs for actual scientists? Perhaps because the unruly scientists fail to behave in accordance with philosophical canons such as explanatory coherence. Machines, on the other hand, can be given predigested evidence and can arrive at the "right" conclusion using the philosophically "correct" method.

Instead of automatically resolving scientific controversies, programs like ECHO could complement detailed historical studies. Freedman (in press), for example, has used ECHO to simulate the latent-learning controversy in psychology. Tolman found that rats learned their way around a maze in the absence of any reinforcement, which apparently contradicted Hull's theory of learning. Indeed, ECHO favors Tolman's explanation over Hull's. But historically, Hull's approach prevailed for many years. Freedman uses ECHO to explore reasons why this might have been; for example, when the weight attached to the latent-learning data is decreased, ECHO delivers a verdict in favor of Hull. Naturally, any hypotheses generated by the computer have to be checked against the historical data; ECHO cannot automatically explain this controversy. Furthermore, ECHO is going to greatly underestimate the role of social negotiations; again, careful historical study can correct for such biases.

9.4. CAN COMPUTATIONAL SIMULATIONS REFUTE SSK?

For at least one philosopher, computer simulations show that scientific discovery and justification can occur without all the messy social negotiations called for by the new sociology of scientific knowledge (SSK). Peter

Slezak (1989) claims that such programs as BACON falsify the new sociology of scientific knowledge, because these programs make discoveries "totally isolated from all social or cultural factors whatever. . . . The claim I wish to advance is that these programs constitute a 'pure' or socially uncontaminated instance of inductive inference, and are capable of autonomously deriving classical scientific laws from the raw observational data" (pp. 563–564). Slezak's claim is consistent with Simon's argument that his programs "discover."

But, as we have seen, that is an ambiguous claim; clearly, many important aspects of discovery lie beyond BACON and KEKADA (see Figure 9-5). The programs do not work from "raw observational data"; instead, their data is sorted neatly into a form designed to facilitate discovery of the correct relationship. Therefore, these programs resemble more what students do in a laboratory exercise than what Kepler did when he struggled to find out what constituted meaningful data and how it ought to be arranged.

Slezak discusses the example of a biology student who replicates some famous experiment. Like Simon, Slezak thinks there is a sense in which the student is a discoverer, in that he or she was previously unaware of the famous result. Simon and Slezak seem ready to argue that every time a student successfully completes a lab he or she has "discovered" a law. But that is a sense of discovery that does not undermine SSK. Kuhn would argue that it is just these sorts of laboratory exercises, coupled with textbook accounts, that give future scientists the exemplars and background knowledge that shape how they see the problem.

Slezak might have cited ECHO as further evidence that an SSK account of scientific progress is unnecessary: ECHO appears to demonstrate that scientific controversies are resolved based on principles like explanatory coherence, not by "black-boxing" and other persuasive strategies favored by SSKers (cf. Latour 1987). But, as we have seen, ECHO merely reproduces the "winner's" arguments, and therefore reflects the results of black-boxing; explanatory coherence itself may be primarily a rhetorical strategy.

9.4.1. CAN SSK BE DISCONFIRMED?

Computers aside, Slezak does raise a valuable question: Is SSK itself refutable, or is it like vulgar Marxism and vulgar psychoanalysis—a theory that explains everything by forbidding nothing? Consider Woolgar's argument that Slezak is treating social influences as extraneous variables, which misrepresents SSK:

> Under this rubric, 'social' no longer refers to extraneous factors which may or may not impinge; instead, it describes the foundational character of all action, thought and behaviour. In particular the action of an individual does not have to take place within a group to be 'social'. Thus, the scientist working on his own is unavoidably part of a language game, in which he envis-

> ages actions, makes sense, interprets and so on, in terms of the conventions which are culturally available. Importantly, 'social' no longer connotes contaminating influences since, in this view, it makes no sense to conceive of the presence of the social as an 'influence', let alone a 'contaminating' influence. (1989:660)

One might counter Woolgar's position by pointing out that, similarly, all phenomena could be construed as cognitive, since they are represented in the mind. Even scientists working in groups must have a mental model or representation of what the group is doing, and this representation is unique: no two individuals see the situation in exactly the same way. Attempts like this to argue that science can be reduced to a single kind of analysis are counterproductive.

Woolgar does sensitize us to the dangers involved in reducing the social—or the cognitive—to a set of variables. This harks back to the problem of translating externally valid findings into ecologically valid settings. Laboratory and computer simulations allow us to isolate variables of critical importance to scientific reasoning, but this artificiality exacts a price. The solution is to maintain a continual dialogue between laboratory and life, as illustrated by the proposed simulation of the canals on Mars controversy (see 8.3) and by a study of inventors in chapter 10.

An SSKer would reject Popper's refutability as a criterion for good theories (cf. Pinch 1985); what constitutes a "good" theory is the product of social negotiations, not rational criteria. SSK, in other words, would presumably prefer to be evaluated on its own criteria, in which case we would wait and see if it succeeded in persuading enough scholars to build a network that would last for generations.

Contra SSK, in this book we have taken what Shadish (1989) calls a "critical multiplist" perspective; theories and approaches must be evaluated by a variety of methods. In the previous chapter we sketched a program of experimental research that would explore the conditions under which groups will adopt and transmit socially negotiated views of a scientific problem or stimulus. These studies cannot "refute" SSK; instead, they can allow us to be more precise about when and where an SSK perspective is most valuable. Naturally, vigorous proponents of SSK would regard this proposition as nonsense, because they "know" that science is simply a product of social negotiations—just as a traditional philosopher "knows" that principles of rationality can account for the advance of science. Both our stereotypic SSKer and our traditional philosopher overgeneralize; each of their perspectives is valuable in certain situations, but not in others.

Future experimental and computational simulations have the potential to help clarify circumstances that encourage social construction. A really successful discovery program, rather that refuting SSK, might help us make externally valid generalizations about the role of different variables in problem choice, representation, and hypothesis evaluation.[23] To make ecologically valid discoveries, it would have to solve what Shrager and Langley (1990) call the problem of embodiment, incorporating "hands-on" craft knowledge and visual as well as propositional representations. It would also have to be able to interact with other programs to form invisible colleges and competing research teams (again, see Shrager and Langley 1990). In fact, it would have to be some kind of sophisticated robot. Even such a robot would not conduct context-independent discoveries; indeed, it would succeed precisely because it could become part of the social negotiations essential to science.

9.5. COMPUTATIONAL METHODS IN PERSPECTIVE

In his review of cognitive science, Gardner argues that one of the greatest benefits of computational simulations has come from their failure to emulate important aspects of human problem-solving:

> But as one moves to more complex and belief-tainted processes such as classifications of ontological domains or judgments concerning rival courses of action, the computational model becomes less adequate. Human beings apparently do not approach these tasks in a manner than can be characterized as logical or rational or that entail step-by-step symbolic processing. Rather, they employ heuristics, strategies, biases, images, and other vague and approximate approaches. The kinds of symbol-manipulation models invoked by Newell, Simon, and others in the first generation of cognitivists do not seem optimal for describing such central human capacities.
>
> The paradox lies in the fact that these insights came about largely through attempts to use computational models and modeling; only through scrupulous adherence to computational thinking could scientists discover the ways in which humans actually differ from the serial digital computer—the von Neumann computer, the model that dominated the thinking of the first generation of cognitive scientists. (Gardner 1985:385)

If Gardner is right, computational simulations may advance technoscience studies by forcing us to be clear about why they are inadequate. Why doesn't BACON discover? Why doesn't ECHO show that explanatory coherence is the key to the oxygen/phlogiston controversy? In other words, computational simulations have negative heuristic value: they force us to be more precise about the role of mental representations, embodiment, and social negotiations. Gardner leaves open the possibility that parallel-processing architectures may succeed where serial computers have failed, and indeed, these sorts of simulations have a potentially important role to play in technoscience studies.

Moscovici (see 5.3) discussed the importance of a consistent, confident style in persuasion. This kind of brazen confidence, coupled with the mystique of a new technology, could cause the potential benefits of a computational approach to be exaggerated. In the language of SSK, what Simon, Thagard, and others are doing is black-boxing a new technique for doing technoscience studies. Those who are suspicious of these conclusions will have to be able to show they can dismantle or unpack the new computational black box.

Like Latour (1987), this book has taken the position that the goal of technoscience studies is to open black boxes. In this spirit, Thagard is willing to send a version of his ECHO program to anyone who asks him for it. However, even possessing a copy of the program does not guarantee that one can "open" it; in fact, as Faraday found when he sent out a packaged version of one of his experiments (Gooding 1990), shipping black boxes to others can be a very effective persuasion technique.

The danger is that we will soon have a computational "strong program" to conflict with SSK. If a dialectic results, then this sort of conflict can advance technoscience studies. But strong programs do not always clash; proponents of SSK and computational cognitivism may simply speak to different audiences, fragmenting science studies. Throughout this book, I have argued that there ought to be collaboration—or at least a dialogue—between those interested in cognitive and sociological approaches to science.

In the next chapter we will consider one attempt to foster such collaboration: preliminary results from a comparative study of inventors that combines cognitive, historical, and sociological perspectives, and may point the way to a new kind of computational approach that complements existing methodologies. As Gooding (1990:167) points out:

> The need for comparative studies of invention and discovery processes is underlined by the shortcomings of computer discovery programs. So far, computational approaches have failed to deliver discovery programs that can work with input that is as "raw" and disordered as is the information that scientists actually work on. This is because many computationalists work with an impoverished notion of discovery drawn from retrospective accounts. The extent of their impoverishment is concealed by the fact that they resemble the idealized accounts that philosophers give. But these in turn are based on highly reconstructed narratives. Programs that purport to model actual discoveries turn out to be based on narratives that are inadequate in at least three respects. One is the predigested and largely symbolic nature of the information recorded in research papers and supplied, perhaps, via a historical account to the program. A second is that situational, context-specific aspects of the process are passed over. A third is that the social, interpersonal dimension so essential to constructing communicable knowledge is largely neglected. Such approaches perpetuate the cognitive and social isolation of the disembodied scientist, once idealized in philosophical theories, and still found in many science textbooks.

TEN

A Cognitive Framework for Understanding Technoscientific Creativity

In this chapter I will develop a cognitive framework that can be used to compare the creative processes of scientists and inventors. At first blush this may seem like a radical departure from the work discussed so far in the book. But in fact such an ambitious project will force a synthesis of, as well as an extension beyond, much of the material discussed in previous chapters.

A great deal has been written and yet very little is understood about creativity (see Perkins 1981 and Hunt 1982 chapter 8 for good overviews). Indeed, there is still no standard definition of creativity.

In this chapter we will skirt the definition issue, and instead try to show how much of the work discussed in the book so far can help us understand and compare the fine-grained cognitive processes of inventors and scientists engaged in acts most of us would agree are creative. As an example, we will consider Alexander Graham Bell's invention of the telephone.

10.1. FRAMEWORKS AND THEORIES

The term *framework* is used here in Tweney's sense:

> I am proposing that we recognize a . . . distinction between claims based on traditional scientific methods, called here *theories*, and claims which attempt to map the complexity of real-world behavior, called here *frameworks*. Truth claims in a theory are based on the familiar strategies of scientific practice, while truth claims in a framework rely on interpretive procedures more akin to the methods of historical scholarship. A theory is an attempt to construct a model of the world which meets certain criteria of testability; it makes predictions, is potentially disconfirmable, and has interesting consequences. A framework is an attempt to *re*-construct a model of the world which meets criteria other than testability as such. An adequate framework is one that is consistent with the details of the process, is interestingly related to our the-

> ories of the world, and reduces the apparent complexity of the real-world process in a way which permits anchoring the framework to the data. In effect, an adequate framework must allow us to see order amid chaos. (Tweney 1989:344)

Tweney's theory/framework distinction can be mapped onto the external/ecological distinction: theories make predictions that can be tested in externally valid experimental situations, whereas frameworks provide a language for organizing and comparing ecologically valid cases. Ideally, interpretative frameworks should be formulated in a way that allows for their potential evolution into theories.

Tweney's own framework for studying and understanding Michael Faraday's cognitive processes was derived, in part, from his own experimental studies of scientific reasoning. It included five levels:

1. Goals and purposes: Tweney assigns factors such as Faraday's religious beliefs to this level, and says that cognitive science has no unique methodological contribution to make to this area, beyond work by such historians as L. Pearce Williams (1965). However, cognitive scientists have made important contributions to our understanding of the role of goals in problem solving (see Wilensky 1983, for example); therefore, any cognitive framework needs to incorporate an inventor's or a scientist's goals at a variety of levels. For example, at the highest level, Faraday sought to unify electricity, magnetism, and gravity; at a lower level, he had specific goals regarding aspects of this unification, e.g., how he was going to translate electricity into magnetism.

2. Cognitive style: Each scientist will differ in the extent to which he or she relies on different cognitive processes, e.g., heuristics, imagery, analogy, and memory; these differences can be studied by cognitive scientists.

3. Heuristics: Tweney defines heuristics as "the strategies which are chosen to organize the path from a starting point to some goal" (Tweney 1989:347). This definition is consistent with our use of the term in this book; previous chapters have provided examples of how heuristics such as confirmation and disconfirmation can be studied.

4. Scripts and schemas: In section 8.3 we discussed the concept of a schema, and related it to Tweney's Faraday research. A script is a procedural schema: it governs the actions a person will take in a particular environment. A classic example is the restaurant script: when one enters a restaurant, one expects to go through a sequence of actions which will produce predictable responses, e.g., when one sits down, one will be brought a menu (cf. Schanck and Abelson 1977).

5. States and operators: Tweney uses these to record Faraday's moment-by-moment activities and thought processes; each mental "state" on a particular problem is followed by a sequence of "operators" which in turn leads to new mental states. These states and operators are derived

from a protocol and allow this protocol to be represented graphically (cf. Ericsson and Simon 1984).

The idea of a "framework" is vague, but that is a strength as well as a weakness. Theories require a level of precision and specificity that is hard to attain early in the study of a complex phenomenon; the danger is that only those parts of the phenomenon which can be fit into a theoretical Procrustean bed will be seriously considered. For example, consider how Tweney's framework might be constrained if it had to be forced immediately into a computational simulation. The highest Level 1 goals would have to be supplied by the programmer; Level 2 would have to emerge from the interaction of Levels 3–5, and even those levels would exclude the visual aspects of mental representation and hands-on skills like tissue-slicing that are vital to experiments. But note that Tweney has been careful to couch his framework in terms that might eventually be computable.

Frameworks allow us to use existing theory to focus and structure our inquiry in a flexible way; we can conduct a fine-grained comparison of the cognitive processes of actual scientists and engineers without forcing ourselves to formulate precise models prematurely.

10.2. A FRAMEWORK FOR STUDYING INVENTORS

Throughout most of this book we have focused on the scientific side of technoscience. But cognitive analyses should be equally useful for technology. What goes on in the mind of the inventor? How do users view a new piece of technology? A thorough study of the inventing mind should lead to changes in our understanding of higher-order cognition.

Most of the psychological observations about inventors have emerged from biographies. For example, Crouch (1989) analyzes the motivations of the Wright brothers from a family therapy perspective. His biography also contains important material about the Wrights' mental processes, but Crouch lacks a framework for interpreting this material. Friedel and Israel (1985) wrote an excellent biography of an invention: Edison's electric light. But while the book is full of important details about Edison's choices, the authors once again lack a framework that would help them draw conclusions about Edison's mental processes.

10.2.1. Brief Biography of a Collaboration

To develop such a framework, I decided to embark on a collaboration with a historian of technology, W. Bernard Carlson. Collaboration is one of the major sources of creativity; if successful, it allows scholars or scientists to step outside standard disciplinary categories and conduct innovative research. Without Carlson's expertise, I could never have tried to apply what I had learned about scientific problem-solving to inventors;

without my expertise, Carlson could never have formulated a framework that would facilitate comparisons between the cognitive processes of different inventors.

Our collaboration grew out of a course that we team-taught at Michigan Tech, called "Innovation and Invention in Science and Technology." The course was a seminar, with Bernie (Carlson) and me debating issues in the history, philosophy, and psychology of science and inviting the students to join in. A side effect was that we learned more than the students—especially about each other's subject areas.

Late one night, on a long drive back from a town one hundred miles to the south, Bernie and I came up with the idea of applying what I knew about the psychology of science to a study of two inventions—the electric light and the telephone. After our first successful grant proposal came through well below our budget, we pared our project down to the study of three telephone inventors: Alexander Graham Bell, Thomas Edison, and Elisha Gray.

A detailed account of this research, funded by the History and Philosophy of Science program of the National Science Foundation, will be published elsewhere (see Gorman and Carlson 1990 for a preliminary report). In this chapter I will outline the framework we developed together and show how it can be applied to Bell's mental processes. Bell is a particularly apt case because he tried to be a scientific inventor, maintaining strong connections with the scientific community (Hounshell 1976; 1981). Therefore, a framework that works with Bell has a good chance of being useful for scientists as well.

10.2.2. Cognition and Invention

Although, as this book has demonstrated, cognitive psychologists are showing an increasing interest in science, there are almost no comparable studies of technology. One of the exceptions is Weber and Perkins (1989), who take the view that invention proceeds by reducing a very large problem-space to a manageable set of alternatives. The classic tool for achieving this sort of reduction is the heuristic. Weber and Perkins, therefore, provide a list of "middle-range" heuristics that ought to facilitate the invention process. By middle-range, they mean that their heuristics lie between the very weak or general and the very strong or domain-specific.

Weber and Dixon (1989) use some of these heuristics to account for the development of the hand sewing needle. One of these is "reduce the time involved in switching between operations." Weber and Dixon conducted an experiment in which an expert and a novice both tried several classic forms of sewing needle. The more modern form of needle led to significantly fewer switching operations than older forms, providing externally valid evidence that the modern needle does greatly reduce switching-time. But Weber and Perkins cannot use this evidence alone to make the ecologically valid claim that switching-time was a driving force in the evolution of the sewing needle; they would have to cite other archaeological and anthropological evidence as well.

One of the major conclusions to emerge from the experimental work described in earlier chapters is the importance of making a distinction between heuristics, which correspond to Tweney's Level 3, and mental representations, which correspond roughly to his Level 4 (scripts and schemas). Subjects can be induced to disconfirm either by altering their mental representations, as in the DAX-MED procedure, or by giving them instructions to disconfirm. But even the instructions, in order to be effective, have to include examples that facilitate an appropriate task representation (see section 3.6).

Representations and heuristics are two components of Tweney's Level 2 (cognitive style) and create a basis for comparing the cognitive processes of individual inventors and scientists, particularly if one takes into account the kind of domain-specific heuristics discussed by Kulkarni and Simon (see section 9.2.2) and the sorts of problem representations described by Larkin and others (see section 8.1.3). Tweney discussed the role of representations and heuristics in Faraday's work (see section 8.1), but there is no comparable study to compare Tweney's with, so one can only draw general conclusions about how Faraday's cognitive style differed from that of his contemporaries.

Weber and Perkins also recognize the importance of the interaction between heuristics and representations. For Weber and Perkins, the key representation is the frame, which they define as "an entity with slots in which particular values, relations, procedures or even other frames reside; as such, the frame is a framework or skeletal structure with places in which to put things" (1989:51). One of their pieces of heuristic advice is simply to represent an invention in terms of a frame with slots; then one can focus on manipulating and combining different slots to create new inventions. Consider the fork: this apparently simple invention can be broken into slots such as "edge" and "grip," and the values of these slots could be varied to create new inventions like the "spork" (a combination of spoon and fork).

Weber and Perkins distinguish between two types of heuristics: between-frame and within-frame. An example of the former is, "Given an artifact or idea that might provide a point of departure for invention, represent it as a frame with slots bound to the values characteristic of the thing or idea" (p. 53). An example of the latter is, "Substitute a variety of values in slots to change the nature of the frame in order to reach a goal" (p. 54).

In the next section we will discuss a specific type of mental representation that is better adapted to studies of the invention process than is a frame, though we will retain the notion of slots as outlined by Weber and Perkins. As these authors are careful to point out, "The frame concept

. . . is by no means necessary. The points we make, for the most part, hold equally for a semantic network, or some other data structure. We note also that simply to represent an invention by means of a frame or another data structure can itself be a significant act of invention" (p. 53).

10.3. MENTAL MODELS

The term *mental model* refers to a particular kind of representation that is especially important in science and technology. Craik (quoted in Tweney, Doherty and Mynatt 1981:326) describes it in this way:

> If the organism carries a "small-scale model" of external reality and of its own possible actions within its head, it is able to try out various alternatives, conclude which is the best of them, react to future situations before they arise, utilise the knowledge of past events in dealing with the present and future, and in every way to react in a much fuller, safer and more competent manner to the emergencies which face it.

There is a long literature on mental models (see Rouse and Morris 1986 for a good review); like so many concepts in cognitive science, the term is used in a variety of not entirely consistent ways. Therefore, it is important to describe some of the most important work in this area. To do this, we will have to detour from our discussion of inventors.

10.3.1. Mental Models in Reasoning

P. N. Johnson-Laird (1983) developed a theory based on mental models to explain how subjects solved syllogisms. Consider the following syllogism:

None of the authors are burglars.
Some of the chefs are burglars.

According to Johnson-Laird, one way the syllogism can be solved is by constructing three mental models, using actors who can take on one or more roles. (Note that one need not actually visualize actors; one can use more abstract symbols in a mental model.) In the first model, one group of actors takes on the role of authors, another of burglars. No actor may be both an author and a burglar. In the second, some of the actors playing burglars assume the role of chef. It is also possible to have some burglars who are not chefs, and some chefs who are not burglars. Six out of twenty subjects run by Johnson-Laird stopped at this point, and concluded that none of the chefs are authors or that none of the authors are chefs. But the correct solution requires constructing a third model, in which all of the author actors also assume chef roles; the syllogism does not prohibit this. Then one has to combine the information from all three models to arrive at the only valid conclusion: some of the chefs are not authors. Seven out of twenty subjects tested by Johnson-Laird managed to reach this conclusion. Six subjects decided there was no valid conclusion.

Johnson-Laird systematically eliminated alternative theories to account for how syllogisms are and ought to be solved, including mental logic, Venn diagrams, and Euler circles. He also simulated his theory of mental models on a computer program that solved syllogisms, and concluded that it was an effective procedure:

> If a procedure can be carried out by a simple machine, plainly it does not require any decisions to be made on the basis of intuition or any other such "magical" ingredient: it is an effective procedure. If a theory is expressed in this form and it is still not obvious what its predictions are, then they in turn should be derivable by an effective procedure from the formulation of the theory. The general criterion, of course, is intended to apply to all scientific theories; it has yet to be satisfied by many psychological hypotheses. Indeed, in contemporary psychology the experimental expertise of a Galileo is often to be found alongside qualitative explanations like those of Aristotle. If the long promised Newtonian revolution in the study of cognition is to occur, then qualitative explanations will have to be abandoned in place of effective procedures. (Johnson-Laird 1983:6)

As noted in 9.1, Johnson-Laird is a proponent of the view that all psychological theories should be expressed in computational form. Johnson-Laird would presumably be uncomfortable with Tweney's notion of a framework because it sounds suspiciously qualitative and "Aristotelian." But one could counter that a framework might be the first step in the articulation of a formal hypothesis, and that trying to develop an "effective procedure" before one really understands a phenomenon qualitatively may be dangerous. Johnson-Laird's effective procedure for solving syllogisms is based only on subjects' final solutions and, in one or two cases, the time it takes them to reach a solution; he provides no details on subjects' processes such as might be provided by protocol analysis. Johnson-Laird needs to provide more qualitative data on the steps subjects use when they try to solve syllogisms. Studies using the 2-4-6 task have certainly focused as much on process as on attaining the experimenter's rule.

To put Johnson-Laird's observation into externally valid language, constructing mental models of each premise ought to be a successful heuristic for solving syllogisms. His experimental evidence certainly supports this claim, but does not warrant the additional ecologically valid claim that people ordinarily use mental models to try to solve syllogisms.

Sections 3.5 and 3.6 emphasized the importance of mental representations on the 2-4-6 task. The kind of mental representation subjects use might be a mental model. In the DAX-MED condition, their model of the task involves two mutually exclusive rules; in the Y-N condition, their model of the task consists of one rule that may have exceptions. Substituting the term *mental model* for *mental representation* does not lead to new insights in the case of the 2-4-6 task, but mental models are a type of representation that are particularly important when we substitute actual scientific and technological systems for abstract laboratory problems.

10.3.2. Mental Models of Technological Systems

It is not clear how Johnson-Laird's conception of a mental model can be applied to complex technological systems. Enter Donald Norman (1983), who has thought about this problem. He distinguishes between:

(1) the target system, which is the technological system the user is trying to understand;

(2) the designer or engineer's conceptual model of how that system works and is supposed to be used;

(3) the user's mental model of the system; and

(4) the cognitive scientist's model of the user's model.

In 3.5 we discussed the relationship between subjects' hypothesized rules and the target rules they were trying to discover. Similarly, Norman's four-part distinction includes the difference between the target system and the user's hypotheses about how to operate it, which are embodied in a mental model. For example, in one study Norman found that people's mental models of the memory stack and registers on a calculator were hazy and incomplete, which meant that they often pushed the clear key and the enter key more times than necessary. In part these key-pressing heuristics evolved from people's experiences with different types of calculators; if one pressed the clear key several times, one could be sure that one had cleared the registers on any type of calculator.

Although Norman does not emphasize it, we see here again the interaction between heuristics and mental models; in this case, instead of trying to form a clear, separate mental model for every type of calculator, some users evolved heuristics that would work across a range of calculators. One of Norman's general conclusions is that people are "willing to trade-off extra physical action for reduced mental actions; they are willing to trade-off extra physical action for reduced mental complexity. This is especially true where the extra actions allow one simplified rule to apply to a variety of devices, thus minimizing the chances for confusion" (1983:8).

Norman lists several characteristics of mental models:

(1) They are incomplete.

(2) They are unstable: people often forget important details of the system they are using.

(3) People's abilities to "run" them are severely limited. In the calculator example above, users had difficulty running their mental models of the memory stack.

(4) Mental models do not have firm boundaries. Again in the calculator example, people's models of different calculators overlapped in a fuzzy way.

Note the way in which Norman emphasizes the limitations of mental models. Cognitive scientists often focus on the constraints, or limitations, of human information-processing systems. But mental models also have powerful strengths, especially for an inventor: they allow her or him to imagine and mentally run new technological ideas, long before it is possible to build a complete working prototype. While novices approaching new devices may have only a limited capacity to "run" mental models of them, expert inventors like Nokola Tesla (Hoffman 1980) can imagine and run a variety of device-like mental representations.

Rouse and Morris argue that "Scientists' conceptualizations of phenomena are almost totally dependent on their own mental models. These models dictate what observations are made and how the resulting data is organized" (1986:359). One could make a similar argument for inventors, as the case of Alexander Graham Bell will illustrate. Rouse and Morris conclude that we can never be sure we have captured an inventor's or user's mental model with absolute accuracy, but that this concept provides a useful tool for exploring and improving system design. Therefore, this concept ought to be useful in developing a framework for comparing the cognitive styles of inventors and scientists.

For example, Wiser and Carey (1983) analyzed a seventeenth-century document called the Saggi di Naturali Esperienze dell' Accademia del Cimento. The "Experimenters," as Wiser and Carey call them, developed a "source-recipient" model of temperature based on the idea that heat and cold are different and are "received" by a body in its natural state from another body which has previously been heated or cooled. This mental model guided their experiments and the interpretation of results.

A mental model may stiffen into one of Lakatos's hard-core assumptions, as this quote from Tycho Brahe illustrates:

> And even if it should appear to some puzzled and rash fellow that the superposed circular movements on the heavens yield sometimes angular or other figures, mostly elongated ones, then it happens accidentally, and reason recoils in horror from this assumption. For one must compose the revolutions of celestial objects definitely from circular motions; otherwise, they could not come back on the same path eternally in equal manner, and an eternal duration would be impossible, not to mention that the orbits would be less simple, and irregular, and unsuitable for scientific treatment. (quoted in Holton 1973:77)

Brahe, like most of his contemporaries, could not imagine orbits being anything but circular. Kepler shared this mental model, but finally rejected it when he could not fit Brahe's data on Mars into a circular orbit:

> The conclusion is quite simply that the planet's path is not a circle—it curves inward on both sides and outward again at opposite ends. Such a curve is called an oval. The orbit is not a circle, but an oval figure. (Kepler, quoted in Koestler 1963:329)

Kepler initially regards this new mental model with disgust: "All he [Kepler] has to say in his own defence is, that having cleared the stable of astronomy of cycles and spirals, he left behind him 'only a single cartful of dung': his oval" (Koestler 1963:329). Eventually, Kepler realized that his oval orbit corresponded to the mathematical ellipse, and his sense of Pythagorean harmony was restored. Kepler had developed a new mental model of the solar system and he was able to translate it effectively into a mathematical formula. Note that this path to discovery is very different from the one taken by BACON, which has no mental models (see 9.2).

Holland et al. describe the development of mental models in a way that will serve as a good summary of this section:

> Although mental models are based in part on static prior knowledge, they are themselves transient, dynamic representations of particular unique situations. . . . Despite their inherently transitory nature—indeed because of it—mental models are the major source of inductive change in long-term knowledge structures. The reason is simple. Because mental models are built by integrating knowledge in novel ways in order to achieve the system's goals, model construction provides the opportunity for new ideas to arise by recombination and as a consequence of model-based predictions. (1986:14)

10.3.3. Expert/Novice Differences in Mental Models

McCloskey (1983) found parallels between medieval views of physical motion and those held by modern novices. For example, a sample of college students held what McCloskey described as an "impetus theory" of motion: an object requires a force to set it in motion, and the force gradually dissipates. McCloskey identifies this view of motion with Philoponus (sixth century) and John Buridan (fourteenth century). Clement (1983) shows that protocols of freshman engineering students working on motion problems resemble Galileo's reasoning in *De Motu*.

McCloskey's and Clement's work suggests that people's mental models of motion are based on everyday experience. For example, when asked to describe the path of a ball dropped by a person walking, most subjects describe it as "straight down" because their experience suggests that the ball lands right by their feet. In fact, the ball describes a parabolic path, moving in parallel with the walker.

Larkin (1983) refers to these sorts of mental models or schema based on objects in the real world as "naive representations"; when confronted with a physics problem, people "run" their naive mental models. In contrast, experts use physical representations, based on such abstract entities as force and motion; these mental models are in turn translated into mathematical representations. Larkin has done extensive studies of experts and novices solving physics problems (see 8.1.3). Her general conclusion is that experts try to translate novel problems into sophisticated physical representations that will, in turn, suggest the appropriate equations; novices, in contrast, have to rely on their naive representations. Larkin does not say whether her representations are equivalent to mental models, but they involve at least some visualization of physical objects and the forces on them. Her research suggests one of the ways in which expert inventors and scientists might formulate and run mental models that are more sophisticated than are models formulated by novices.

10.3.4. Instructions to Use Mental Models

Gentner and Gentner (1983) conducted a series of experiments in which they explored how mental models affected subjects' abilities to solve electrical problems. In one study, subjects were taught to use either a "moving crowd" or a "flowing waters" model of electricity. In the former, forward pressure on the crowd, corresponding to voltage, was created by a loudspeaker shouting encouragement; in the latter, reservoirs or pumps created pressure. The moving-crowd model led subjects to a better understanding of resistors, particularly parallel resistors, than did the flowing-waters model. The authors predicted the reverse pattern for multiple battery systems, but this prediction was not confirmed, primarily because subjects in the flowing-waters condition did not understand the way water would behave if there were multiple parallel reservoirs or pumps. An earlier study showed that subjects who naturally used flowing waters as an analogy for electric current did much better on battery problems than on resistor problems.

Like the study by Gorman, Stafford, and Gorman reported in 3.4, the Gentner and Gentner study provides externally valid evidence that changing subjects' mental models affects their performance on technoscientific problems. It also makes an important point about analogies as potential sources of mental models.[24]

10.3.5. Bell's Mental Model

Ferguson reports the example of an electrical engineer who argued that

> "Creative individuals appear to have stumbled onto and then developed to a high degree of perfection an unusual ability to visualize mentally—almost hallucinate—in the areas in which they are creative." He goes on to explain that, to him, Ohm's law is not merely an equation, but a "vision and feeling about something like a fluid stuff, trying to flow through a solid stuff that opposes the flow, and . . . the harder the electrical stuff is pushed, the more rapidly it flows through the resistance opposing its movement." He builds in his mind "a vivid and manipulative image system of the flow of electrical

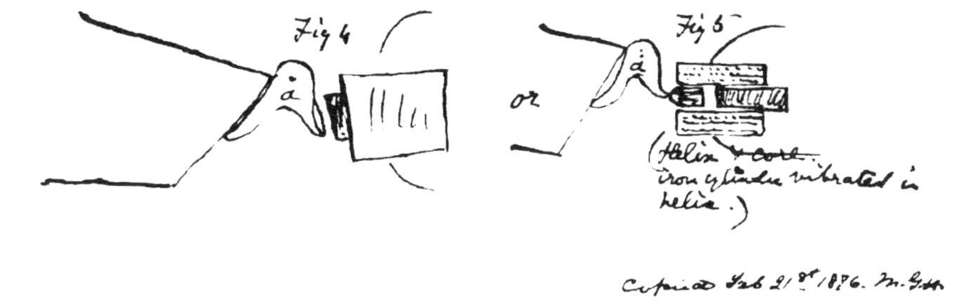

FIGURE 10-1 BELL'S EAR MENTAL MODEL
"Figure 4," from left to right, shows a cone and diaphragm with the bones of the ear (a) attached. As one speaks into the cone, the bones vibrate in front of an electromagnet which connects mechanical motion into electric current. "Figure 5" shows a different electromechanical arrangement next to the ossicles. The text under it reads "Helix & core. iron cylinder vibrated in helix." (From "Experiments made by A. Graham Bell, Vol. I" notebook, p. 13, Box 258, Bell Family Papers, Library of Congress.)

currents" that permits him "to perform a myriad of mental experiments in a very short time." (1980:12–13)

Similarly, I would argue that a key component of Alexander Graham Bell's creativity was his ability to manipulate mental images, going back and forth between his experiments and his evolving mental model of what he was trying to achieve. In his notebooks, he gives us occasional glimpses of these mental models.

Consider, for example, his entry of February 21, 1876 (see Figure 10-1). Bell has sketched the bones of the inner ear vibrating against an electromagnet. As one spoke into the cone at the left, these bones would vibrate, and the vibrations would be converted into electricity and transmitted to a similar device. Obviously, the bones of the inner ear cannot be used in this way; what Bell is doing is sketching his mental model, a device he can run in his imagination.

The text for this entry reads, "Make transmitting instrument after the model of the human ear. Make armature (a) the shape of the ossicles. Follow out the analogy of nature" (Bell 1876b:13). The "transmitting instrument" here is the transmitter for a telephone. Bell has told us he will base his telephone on an analogy with the human ear. But he has done more than simply state this analogy. He has sketched the mental model he has derived from it, showing us an apparatus that could never be built, only imagined—a device in which the bones of the ear transmit the vibrations from a diaphragm reminiscent of the eardrum to two different electromagnetic arrangements. These arrangements are examples of what we call mechanical representations: familiar devices or solutions an inventor can put to use when imagining how his or her mental model might run. (We will discuss mechanical representations below, using the origin of the left-hand arrangement in Figure 10-1 as an example.)

Bell did not originally set out to build a telephone; instead, he joined the chase for the multiple telegraph, the cutting-edge technology of his day: it would permit businesses to send multiple messages simultaneously over the same telegraph wire. But, as David Hounshell (1981) has pointed out, Bell was an outsider to the telegraph community and his devices were more amateurish than were similar devices constructed by other telegraph inventors such as Elisha Gray and Thomas Edison.

However, Bell also had a unique area of expertise. He was a teacher of the deaf who knew a great deal about how sound waves were transmitted by the vocal chords and received by the ear. It was this background that made him appreciate the importance of transmitting spoken communication.

Indeed, he had a long-standing interest in making speech visible to the deaf, so a device called a phonautograph particularly interested him: one sang or spoke into a wooden cone, and a diaphragm at the end of the cone with a bristle brush attached to it traced the shape of the wave onto a piece of smoked glass. In 1874, at the instigation of Clarence Blake (1875), Bell constructed a phonautograph using an actual human ear as the speaking tube and the bones of the inner ear as the connection between the eardrum and the bristle brush (see Figure 10-2). This experiment gave Bell a visuo-tactile understanding of how speech could be translated into an undulating mechanical motion; this understanding, in turn, was incorporated into his ear mental model, where electromagnets are substituted for the bristle brush.

So, Bell's hands-on experience with devices like the phonautograph gave him a unique expertise that helped him see the potential for a device that could reproduce human speech. But Bell had to develop the skills and knowledge necessary to reproduce sounds electrically. The great psychophysicist Helmholtz had demonstrated that vowel sounds could be produced by an electromechanical apparatus involving resonators, tuning forks, and electromagnets (Bruce 1973; Rhodes 1929). Bell saw this apparatus, and it provided him with an idea for a multiple telegraph: why not use multiple tuning forks to send several messages simultaneously over a single wire? Each of the sending forks could be tuned in precise harmony with one of the receiving ones; simultaneous messages would not interfere, because each would be sent and decoded separately by one of the pairs of tuning forks.

Bell spent a great deal of time trying to transmit messages using tuning forks. In the course of these experiments, he gradually developed a new mechanical representation which would later appear in his sketch of his ear mental model.

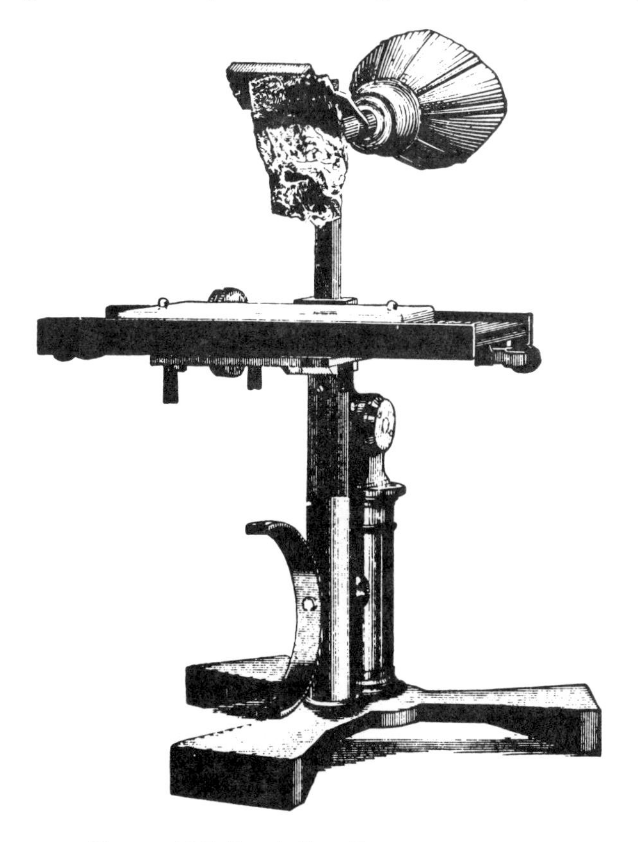

FIGURE 10-2 BELL'S EAR PHONAUTOGRAPH
The bones of the ear are mounted over a piece of smoked glass; as one speaks into the ear-shaped cone at the top, the undulating sound waves are traced on the glass. (From Prescott 1884:82.)

10.4. MECHANICAL REPRESENTATIONS

W. Bernard Carlson and I developed "mechanical representations" as a component of cognitive style, to be applied specifically to the case of invention. The closest analogy we have been able to find is De Kleer and Brown's (1983) concept of a mechanistic mental model. These authors discuss how a mental model of the mechanical workings of a device must be constructed in order to understand its functioning. This overall mechanistic model may contain sub-models for specific components of the device. It is also possible for a troubleshooter to consider several different mechanistic models of a device, much as a scientist can consider several alternative hypotheses. In general, the expert is more aware of his or her overall mechanistic model and the assumptions on which it is based than is the novice, whose models and assumptions are more likely to be tacit.

One implication of this view is that an expert troubleshooter can test and discard alternative mental models, much as a scientist can test and discard alternative hypotheses. Future research could explore ways in which troubleshooters try to confirm or disconfirm their mechanistic mental models.

De Kleer and Brown's troubleshooter is trying to develop an overall mental model of a malfunctioning device that may include lower-order mental models of specific components. Inventors have to possess troubleshooting skills; they have to be able to tell why a device they have built is not working. But they have to be able to imagine and construct novel devices. One of the ways of reducing the complexity of this task is to rely on familiar devices for components—mechanical solutions that the inventor has relied on repeatedly in the past to make inventions work.

We refer to these familiar mechanical solutions as "mechanical representations," to denote the fact that the inventor can visualize them as components in his or her "mind's eye" and run them as part of an overall mental model of a new invention. For example, Edison took the drum cylinder he had used on the phonograph and covered it with a photographic emulsion to create one of his early kinetoscopes (see Carlson and Gorman 1990). As Reese V. Jenkins says (1984:153), "Any creative technologist possesses a mental set of stock solutions from which he draws in addressing problems." This "mental set of stock solutions" is what we mean by mechanical representations and serves as part of what Giere (1988) calls a scientist's or inventor's cognitive resources.

Whereas mental models are often "incomplete," "unstable," and "do not have firm boundaries" (Norman 1983:8), a mechanical representation is clear and precise: it is the mental representation of an operable component of a device. One can identify mechanical representations by noting their repeated use in inventions. Some mechanical representations are unique to individual inventors; others are shared. Ferguson (1977) has described "picture books" that contained hundreds of what we are calling mechanical representations; inventors could freely borrow solutions from these books. He further argues that much of our modern mechanical knowledge has a long history; he cites the fact that almost all of our modern forms of gears were included in a book compiled in 1588 by Ramelli. Mental models of new inventions can arise from creative tinkering with mechanical representations. As Robert Fulton said, "the mechanic should sit down among levers, screws, wedges, wheels, etc., like a poet among the letters of the alphabet, considering them as the exhibition of his thoughts in which a new arrangement transmits a new idea to the world" (quoted in Philip 1985:47).

Inventors freely borrow mechanical representations from each other, as well as from the machine books cited by Ferguson, while trying to patent their own novel mechanical solutions.

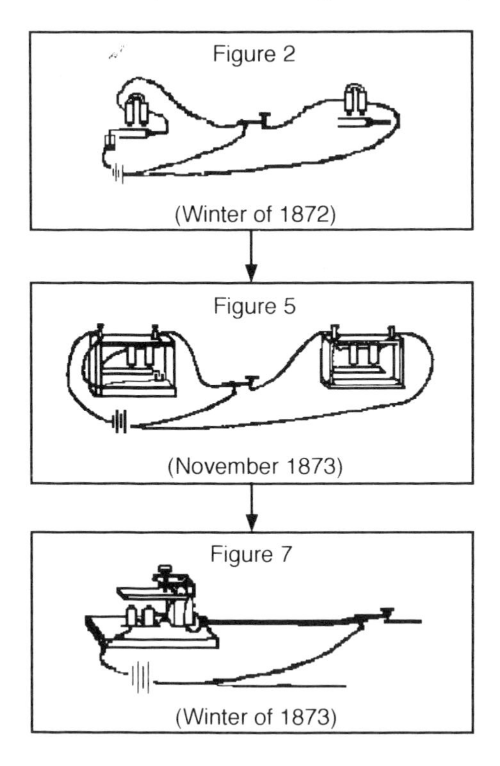

FIGURE 10-3 FROM TUNING FORK TO REED RELAY
The tuning forks in the top figure gradually evolve into steel plates suspended under the magnets in the middle figure and a plate over the magnets at the bottom, where the mercury cup has also been omitted. (These figures are based on the corresponding figures in Bell, 1876a.)

10.4.1. BELL'S MECHANICAL REPRESENTATIONS

When Bell began his first multiple telegraph experiments, the combination of tuning fork and electromagnet, borrowed from Helmholtz, served as a mechanical representation. Figure 10-3 illustrates how Bell evolved a new mechanical representation from this tuning fork. The top diagram shows the connection between transmitting and receiving tuning forks. This combination is then replaced by a steel plate or reed suspended below an electromagnet. Finally, the steel reed is inverted, the box or chamber in which it was housed is removed, and an adjusting screw is added at the top. Bell used this mechanical representation frequently in his inventions, and gave it a conspicuous role in his telephone patent (see bottom, Figure 10-3).

Let us explore how a new mechanical representation affects the con-

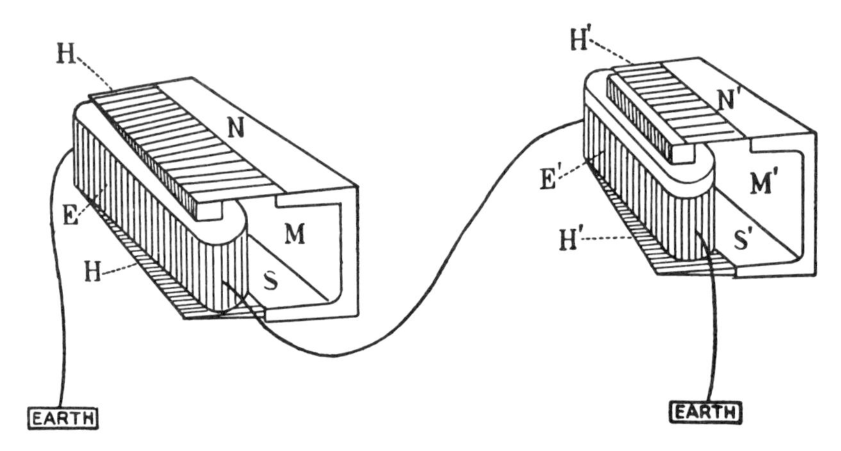

FIGURE 10-4 THE HARP APPARATUS
As one spoke into N, the reeds H would vibrate over the electromagnet E; these vibrations would be transmitted no N'. Bell sketched, but never built, this apparatus in the summer of 1874. (Drawing is from Rhodes 1929:11.)

struction of subsequent mental models. In 1874, Bell sketched, but never built, a harp apparatus that consisted of a row of steel reeds suspended over a single, large electromagnet (see Fig. 10-4). As one spoke a vowel into one of Bell's "harps," the reeds would vibrate, setting up what Bell called an "undulating current" that would be transmitted to the receiving harp. Bell imagined this undulating current having the form and shape of the undulations produced by the phonautograph when it traced sound waves on smoked glass.

Bell's harp represented a potential improvement over the first "telephone," developed by Philipp Reis in Germany in 1861. Reis's device consisted of a membrane diaphragm connected to a needle that dipped in and out of a small cup of mercury; as one sang into the diaphragm, the needle made or broke contact with the mercury, producing a kind of intermittent or on-off current.

The Reis telephone reproduced only the fundamental tone or pitch of a sound; the subtle overtones and their corresponding amplitudes were missing. In contrast, one of the reeds in Bell's harp apparatus would transmit the fundamental tone; others, vibrating in sympathy, would pick up the feebler overtones; the receiving harp would faithfully reproduce every nuance of a vowel sound spoken into the harp.

Bell never built this device; he was not sure how many reeds it would take to reproduce the human voice, and he thought the induced currents would be too weak. Instead, the harp served as a mental model of how sound waves might be translated into an undulating electric current.

Bell and his redoubtable assistant Watson tried through the winter and spring of 1875 to develop a harmonic multiple telegraph; Bell's backers, including his father-in-law, wanted him to focus on this practical problem and not waste his time chasing the idea of a speaking telegraph. Whereas in the harp all the reeds were mounted over a single electromagnet, in a multiple telegraph each tuned reed would be on a separate electromagnent; these reeds would replace the tuning forks used in Bell's earlier harmonic multiple telegraph and would be arranged so that each reed would be paired with a partner tuned similarly, allowing messages from four or more reeds to be sent simultaneously over a single wire.

For this apparatus to work, the reeds on sending and receiving ends had to be tuned very precisely. On June 2, 1875, one of the many problems that arose with this pesky apparatus taught Bell an important lesson. He and Watson were tuning the transmitter and receiver reeds when one of the transmitter reeds on Watson's end became stuck against the magnet. Watson was plucking it to set it vibrating, when Bell burst into the room (Watson 1913:10–11):

> "What did you do then? Don't change anything. Let me see!" I showed him. . . . The contact screw was screwed down so far that it made permanent contact with the spring, so that when I snapped the spring the circuit had remained unbroken while that strip of magnetized steel by its vibration over the pole of its magnet was generating that marvelous conception of Bell's—a current of electricity that varied in intensity precisely as the air was varying in density within hearing distance of that spring. That undulatory current had passed through the connecting wire to the distant receiver which, fortunately, was a mechanism that could transform that current back into an extremely faint echo of the sound of the vibrating spring that had generated it, but what was still more fortunate, the right man had that mechanism at his ear during that fleeting moment, and instantly recognized the transcendent importance of that faint sound thus electrically transmitted.

When this bit of serendipity occurred, it showed Bell that his harp apparatus mental model was unnecessarily complex. A single reed could reproduce speech as faithfully as would the collection of reeds in the harp; furthermore, it was possible that this induced undulating current could be made powerful enough for long-distance transmission. Bell's phonautograph experiments had shown him that a vibrating diaphragm could reproduce sound waves; this reed experiment indicated that these mechanical vibrations could be translated into an undulating electrical current by a single reed relay hinged on the diaphragm in a manner similar to the arrangement of the bones of the human ear. Now all the elements of his ear mental model were in place.

That same night, Bell sketched what came to be called his "Gallows telephone"; Watson built it for him the next day (see Figure 10-5). When one spoke into the mouthpiece of this instrument, the diaphragm vibrated, causing the reed to vibrate and transmit an undulating current to

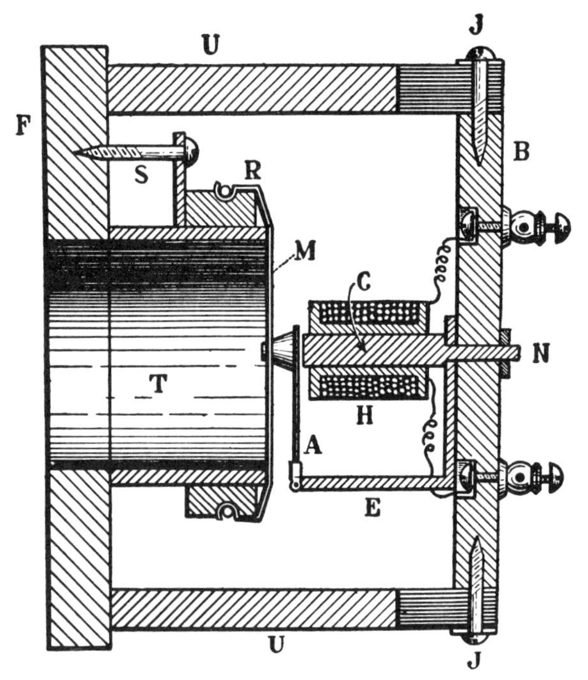

FIGURE 10-5 THE GALLOWS TELEPHONE
Bell's "Gallows Telephone" of June 1875. One speaks into T, which causes the membrane M to vibrate; these vibrations are transmitted to the Armature A, which in turn induces a current in the coil H. (Drawing is from Rhodes 1929:24.)

a receiving reed. This device was, in effect, a kind of electromechanical ear.

Bell's Gallows telephone showed how a voice could theoretically be transmitted electrically. But this instrument did not work very well; what one heard on the other end was an indistinct mumbling. Still, Bell felt he had discovered a means to transmit speech electrically. In January 1876, Hubbard filed a patent on Bell's behalf. Just after filing this application, Bell stated his mental model in his notebook (see Figure 10-1).

To summarize, the development of a new mechanical representation affected the evolution of Bell's mental model of the telephone. The substitution of reeds for tuning forks in Bell's early telegraph experiments played a critical role in the harp apparatus and in his Gallows telephone: it is hard to imagine either of these devices being constructed with tuning forks.

10.5. HEURISTICS

The evolution of Bell's ear mental model suggests the importance of heuristics as well. Bell stated he was "following the analogy of nature." This is a heuristic. (We have discussed heuristics extensively in earlier chapters; see, for example, 2.3.1, 5.1.2, and 9.1.1. Developing this framework forced me to confront the notion of a heuristic once again, so I include here a brief discussion that will give additional background necessary for understanding the interplay of heuristics and mental models.)

According to Groner, Groner, and Bischof (1983) the term *heuristic* is derived from the ancient Greek verb *heuriskein*, which means to find or discover. A heuristic is a general strategy or "rule of thumb" which guides the discovery process. Nisbett, Krantz, Jepson, and Fong describe a heuristic as "any guiding principle for transforming information to solve a problem or form a judgment" (1982:447).[25]

George Polya "is recognized in AI as the person who put heuristic back on the map of intellectual concerns" (Newell 1983:195). His heuristics included *reductio ad absurdum*, working backwards, going back to definitions, and accumulating knowledge. One reason expert systems are so difficult to set up is that "the expert must communicate not merely the 'facts' of his field but also the heuristics: the informal judgemental rules that guide him. These are rarely thought about concretely, and almost never appear in journal articles, textbooks, or university courses" (Lenat 1983:352).

So, when Bell said he was "following the analogy of nature," he was stating an informal rule of thumb or principle that guided his work. One can see evidence of this heuristic as early as 1866, when Bell and his brother Melville—on his father's advice—built a gutta-percha model of the human vocal apparatus. They were actually able to produce speech sounds with their artificial mouth and throat (see Bruce 1973:35–37 for details). Bell's preference for the "follow the analogy of nature" heuristic may date from this experience. In addition to prompting him to base the telephone on the ear, it may in part explain his obsession with transmitting an undulating current: he wanted to send an electric signal that looked just like a sound wave.

The application of the "analogy of nature" heuristic depends critically on having an appropriate mental model. Consider what would have happened if Bell's competitors, Elisha Gray or Thomas Edison, had tried to employ this heuristic. They lacked Bell's extensive, hands-on knowledge of speech and hearing and could not have formulated an appropriate mental model based on following the analogy of nature.

10.5.1. Heuristics and Mental Models in the Mountains

In 3.6 we discussed the relationship between heuristics and representations. Let me include a personal example here that will illustrate the interaction between mental models and heuristics. I climbed Mount Olympus behind Salt Lake City on a day when mist fell over the summit and obscured all but the rocks within a few feet. I knew I would have to rely on my fuzzy, incomplete, three-dimensional model of the contours of the peak to find my way back. After about fifteen minutes of hard scrambling, there was a mismatch between my model of where I was on the peak and the surrounding terrain. I had a friend posted below; if I were on the right course, he would be able to hear me. I shouted; no response.

This "experiment" had disconfirmed my mental model of my position. Now I wasn't sure where I was in relation to my goal, but I had been careful to mark my route back to the summit. So I relied on a "weak" heuristic: when you've lost the way, go back to the last place where you knew your location. When I reached the summit, I carefully experimented with different routes down, retracing my steps if they didn't seem promising, until I found one that worked, as confirmed by my friend's answering shout.

This example illustrates the importance of heuristics when models fail. But even my "retracing steps" heuristic depended on a mental model indicating where I had been, and memory for specific cues. If I had not carefully stored that information, then I would have had to rely on an even weaker and more general heuristic—stay put until someone finds you. In cold and mist, this heuristic can lead to hypothermia, but it is preferable to falling off a cliff.

10.5.2. Mental Models and Algorithms

In 3.6.2, we discussed how heuristics could be arranged in a a hierarchy, from general to specific. Lower-level, domain-specific heuristics are commonly referred to as algorithms: "step-by-step procedures that 'mechanically' produce a solution to any problem out of a certain class of problems" (Groner, Groner and Bischof 1983). More general heuristics are useful in situations where there are no algorithms (Dorner 1983). Algorithms are guaranteed to produce a solution in a finite number of steps (Haugeland 1986).

But there is no guarantee that this solution will be appropriate to the problem at hand. Experts in a particular domain possess a large repertoire of algorithms to complement their declarative knowledge. Mental models or representations based on this knowledge suggest when a particular algorithm is appropriate. It is possible for the mental model to suggest an inappropriate algorithm. Donald Norman's *The Psychology of Everyday Things* (1988) is full of examples of devices that cue inappropriate mental models which in turn suggest algorithms that do not work. Consider the case of a colleague of Norman's who had trouble with the turn-signal switch on his motorcycle. Moving the switch forward signaled a right turn and backward a left—but the switch was mounted on the left handlebar, and so the colleague developed an inappropriate mental model of the action: if you push a switch on the left handlebar forward, you should signal a left turn. To learn the proper motion, he had to develop a new mental model, in which the key element was pushing the switch the same way you would turn the left handlebar—forward for a right turn, backward for a left.

I have had a similar experience teaching novices how to use a statistics package on a mainframe computer. To explain how the system worked, I used an analogy to a word processor, with which most of these students were familiar; in effect, I helped them create a mental model in which their user area was like a disk and the statistics program had to be accessed via the operating system, just like most word-processing programs. But for those who had no experience with word processors, developing a mental model was quite difficult and led to errors like trying to type statistics commands when they were in the operating system and vice versa. Even those students who mastered the commands would apply them at incorrect times until they developed an appropriate mental model. Once again, heuristics and procedures depend on mental models.

Often, the expert is operating in a somewhat ambiguous situation. Consider the case of the "troubleshooter" (De Kleer and Brown 1983). Declarative knowledge provides appropriate models and algorithms for virtually instantaneous diagnosis on familiar problems, but in unfamiliar cases the troubleshooter will have to follow the heuristic of imagining how the device is functioning and which components must have broken down to cause the problem. If the expert constructs an inappropriate mental model, then even the most reliable algorithms will not repair the device.

10.6. MENTAL MODELS AND HYPOTHESES

Much of the research in chapters 1 through 6 concerned the role of heuristics such as confirmation and disconfirmation in the process of theory formation and evaluation. What role should theory play in a cognitive framework that hopes to improve our understanding of invention and discovery?

Let us consider Bell again. As Hounshell (1976) has shown, Bell identified closely with the scientific community and therefore tried to be as scientific as possible in his work. Edison and Gray, in contrast, were not concerned about being scientific; their goal was to build working devices by any method that produced promising results. When Edison needed scientific expertise, he hired someone who had it.

Bell tried to develop hypotheses in the course of his invention. For example, he noted that when an intermittent current was passed through an induction coil, the coil emitted a sound similar to that of the receiver in the Reis telephone. Bell demonstrated his discovery to one of America's most distinguished scientists, Joseph Henry, who advised Bell to work at this "germ of a great invention." When Bell admitted that he lacked the necessary electrical knowledge, Henry told him to "GET IT!" (Bell, letter to "Papa and Mamma," March 18, 1875—see Bell 1908:53).

In a May 4, 1875 letter to Gardiner Hubbard, his father-in-law and backer, Bell wrote, "The deduction I had made from the experiments was that an intermittent current of electricity creates a molecular vibration in the conductor through which it is passed. Prof. [*sic*] Henry writes that he believes my experiments are new, and that my explanation of the cause of the sound is in his opinion correct" (cf. Bell 1908:53).

Bell's theory included the idea that increasing the "molecular tension" of a wire would increase its resistance. A vibrating wire on a piano or violin would experience rapid changes in tension; therefore, a continuous current passing through such a wire would meet with a varying resistance. The vibrations of the string should cause corresponding oscillations in the current at the same amplitude and frequency, allowing the timbre of the tone to be transmitted.

Bell did propose an experiment to confirm or disconfirm his molecular vibration theory. He intended to pass an intermittent current through one of the strings of a neighbor's piano; if his prediction were right, molecular vibrations would cause the string to sound. Unfortunately, the experiment failed because of the way the strings were attached to the metal frame. In fact, Elisha Gray was able to successfully send tones from pianos and violins in his pursuit of a harmonic multiple telegraph.

Bell's molecular vibration theory provided him with an alternative mental model. As he said in his letter of May 4 to Hubbard, "The plan for transmitting timbre, that I explained to you before,—viz. causing permanent magnets to vibrate in front of electromagnets,—is chiefly deficient on account of the feebleness of the induced currents. If the other plan is successful, the strength of the currents can be increased ad libitum without destroying the relative intensities of the vibrations" (Bell 1908:53). This earlier plan was based on the reed relay and led to Bell's Gallows telephone. The "other plan" was this idea of substituting vibrating strings for reeds or tuning forks.

It was not until 1879 that Bell tried an experiment in which he passed a continuous current through a thin steel wire. The sound of the vibrating wire was picked up clearly by one of his commercial telephone receivers in the same circuit.

Why did Bell wait so long to confirm this theory? Because the June 2 experiment with Watson showed him that the currents induced by the reeds would be sufficient to transmit tones and speech. In other words, the June 2 result provided Bell with an important confirmation during the pursuit phase of inquiry, and oriented him toward the magneto rather than the vibrating string mental model.

What we do not know is whether Bell's molecular vibration hypothesis preceded or was derived from his mental model. Giere (1988), following Kuhn's idea of an exemplar, identifies theories as clusters of mental models:

> As the ordinary meaning of the word 'model' suggests, theoretical models are intended to be models of something, and not merely exemplers to be

used in the construction of other theoretical models. I suggest that they function as 'representations' in one of the more general senses now current in cognitive psychology. Theoretical models are the means by which scientists represent the world—both to themselves and for others. They are used to represent the diverse systems found in the real world: springs and pendulums, projectiles and planets, violin strings and drum heads. . . . Unlike models, a theoretical hypothesis is, on my account, a linguistic entity, namely, a statement asserting some sort of relationship between a model and a designated real system (or class of real systems). (pp. 117–118)

Giere's use of the term *theoretical model* is not necessarily identical to our use of the term *mental model*. But Giere's examples—pendulums, projectiles, and planets—suggest a form of device-like representation akin to a mental model. So one might derive from Giere the argument that mental models are nonlinguistic representations that precede and underlie theories.[26]

In contrast, Holland, Holyoak, Nisbett, and Thagard seem to argue that "theories can be understood as systems of rules furnishing mental models" (1986:327), which implies that mental models are derived from theories: "A high-level theory such as Newtonian mechanics or Darwin's theory of evolution organizes our thought processes in fundamental ways, shaping our approach to every problem that might fall under it. A new theory thus brings with it not only different declarative knowledge but different types of problem solutions and different patterns of activation of concepts" (Holland et al. 1986:331).

A simple synthesis would be to argue that mental models can either lead to, or be derived from, hypotheses. Inventors rarely state theories explicitly; to explore the relationship between mental models and theories, this framework should be used to compare scientists working on the same or similar problems.

Gooding (1990) has taken an important step in this direction by showing how mental models, heuristics, and mechanical representations can be used to shed light on Faraday's discovery of electromagnetic rotation, which in turn led to the invention of the first induction motor. Interestingly, Gooding concludes that the traditional "hypothesis/conjecture-test" approach "lacks the resources needed to describe the interplay of thought and action" (186). He particularly emphasizes that "It can take longer to achieve uniform performance of a real device than to achieve a reliably self-consistent mental representation of such a device" (189). Like Bell, Faraday lived in a world where both hypotheses and mental models were often easier to create than were reliable, working devices.

10.7. SLOTS IN MENTAL MODELS

In fact, at the time Bell completed his telephone patent, he did not have a working device. Immediately afterward, he launched into a long series of experiments in which he tried a different principle for the telephone:

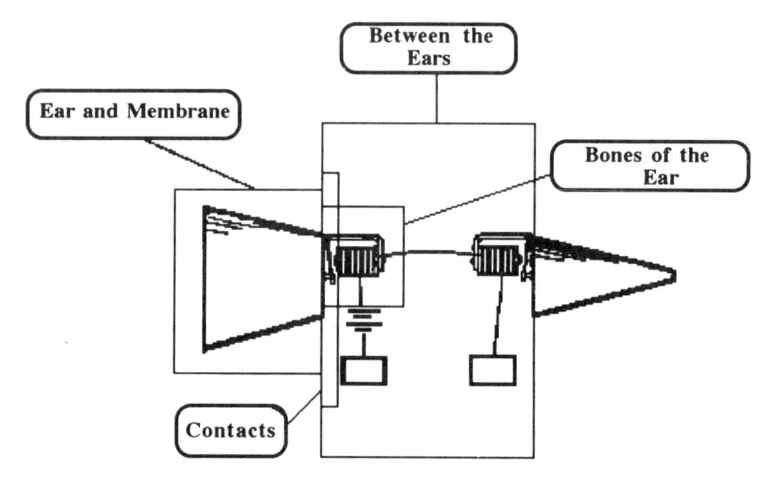

FIGURE 10-6 SLOTS IN BELL'S EAR MENTAL MODEL
The underlying drawing is from Bell's patents "Improvement in Telegraphy," 174, 465 (executed 20 January 1876, granted 7 March 1876).

adding a variable resistance medium between transmitter and receiver. To understand Bell's subsequent experiments, we will have to introduce another aspect of our framework.

To clarify the relationship between mental models and mechanical representations, W. Bernard Carlson and I have modified Weber and Perkins's notion of a slot (see 10.2.2). In our view, inventors' mental models can be divided into slots into which mechanical representations can be inserted. Figure 10-6 shows how Bell's ear mental model can be translated into a slot diagram. The underlying drawing comes from Bell's successful patent application. Although Bell did not have a working device when he filed his patent, he had a mental model of how speech could be transmitted electronically, and he succeeded in patenting that idea. So his patent sketch reflects only one possible way of building a successful telephone.

To illustrate the potential for alternatives, we have divided this sketch into slots, based on Bell's analogy between the telephone and the ear. There is, for example, a slot corresponding to the bones of the ear, in which Bell inserted different mechanical representations, such as the reed relay and the helical coil depicted in Figure 10-1. Bell also used different combinations of cones and diaphragms in the "ear and membrane" slot. He tried different arrangements for improving the line transmission "between the ears." Finally, in his experiments immediately after the patent, he opened a new slot corresponding to the contact between diaphragm and reed relay, into which he inserted water as a variable resistance medium (cf. Finn 1966). This, in turn, led to his first successful transmission of speech.

Weber and Perkins typically describe slots in terms of lists of functional attributes. If we wished to make our analysis consistent with theirs, we would describe Bell's "bones of the ear" slot in terms of the actions he wished to accomplish, e.g., produce undulations. But we think that Bell's "bones" slot was more than a list of functions: he wanted to find a set of mechanical representations that would do just what the ossicles did. This helps explain why he abandoned other methods of producing undulations to come back to ones that came closer to mimicking the bones of the ear. Telegraph experts like Gray and Edison were sure his magneto telephones would never work very well, but to Bell they were a simple, direct realization of a mental model that worked well in nature. Why shouldn't a mechanical ossicle outperform more complicated alternatives?

Shortly after his patent, Bell began a series of experiments in his ossicles or "bones of the ear" slot. He started with an experimental configuration similar to his earlier Helmholtz apparatus (see Figure 10-7). The familiar tuning fork is placed over an induction coil, and a reed relay is used as the receiver. To improve the sound produced by this device, Bell gradually made a series of substitutions, using the holding-constant heuristic to focus on one variation at a time. Figure 10-7 represents an oversimplification of this process, showing only two of these substitutions. He removes the electromagnet below the tuning fork; this circuit produces unsatisfactory results, so he inserts a dish of water.

Note that the former location of the electromagnet is now serving as a kind of subslot under the ossicles slot. We have labeled this slot "contacts" because Bell focuses on the relative sizes of the contacts separated by the water.

The bottom of Figure 10-7 shows the eventual result, which produced the first successful transmission of speech: "Mr. Watson—Come here—I want to see you" (Bell 1876b: Vol. I, p. 40).

10.8. A TALE OF THREE INVENTORS

On the same day (February 14, 1876) that Bell's application arrived at the patent office, Elisha Gray submitted a caveat based on the liquid variable resistance principle. (A caveat is a notice of intention to file a patent—unlike Bell, Gray hesitated to file an actual patent until he had a working device.) Lloyd Taylor, Gray's unpublished biographer, uses the fact that Bell's first successful transmitter resembles the device sketched in this caveat to make the claim that Gray deserves the real credit for inventing the telephone because he came up with the idea first.

Who gets credit for a discovery or invention is an interesting sociological question that lies beyond a cognitive framework. Gardiner Hubbard, Bell's father-in-law and principal backer, was the one who filed the patent and who promoted his son-in-law's cause through a long series of patent litigations. Bell himself was an excellent patent witness who gave convincing and eloquent reconstructions of his path to invention; in contrast, the

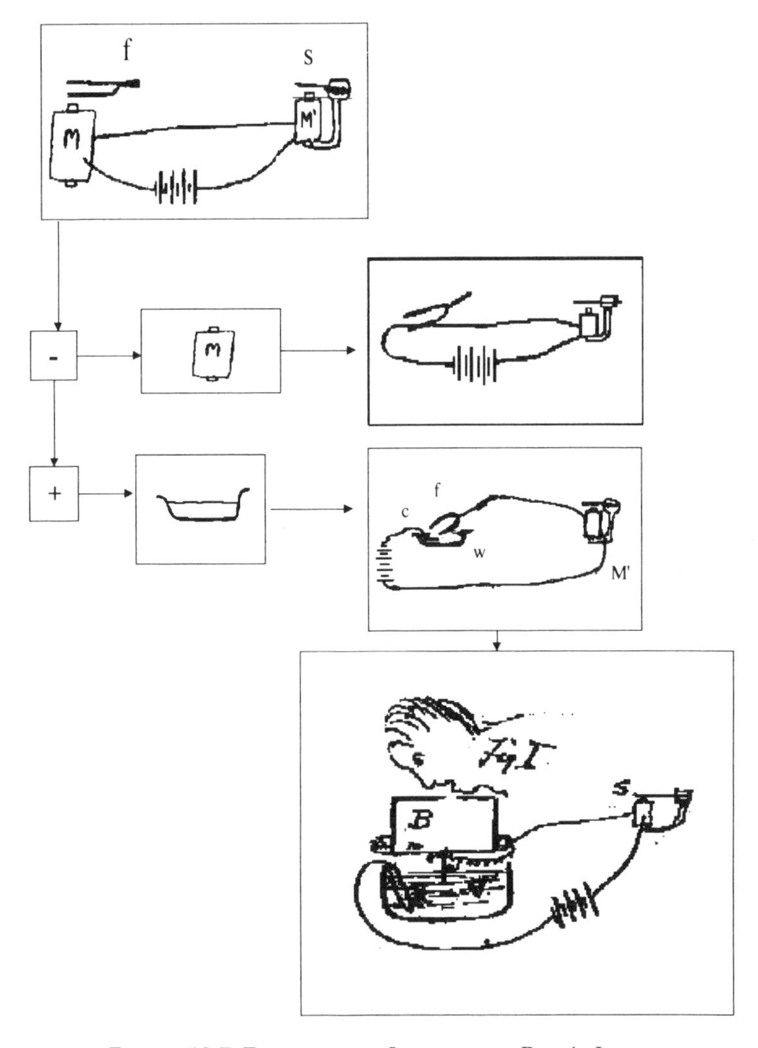

FIGURE 10-7 EXPERIMENTS LEADING TO BELL'S LIQUID TRANSMITTER
This figure shows two of the major substitutions Bell made within the "bones of the ear" slot. The device on the bottom illustrates his first successful transmission of speech, on March 10, 1876. All sketches are from Volume I of "Experiments Made by A. Graham Bell," pp. 35, 37, & 39 (Library of Congress).

relatively more taciturn Gray had trouble remembering important details and had lost much of the evidence that might have supported his case. Gray also wrote letters to Bell in which he gave the latter credit for inventing the telephone—letters Gray later regretted.[27] So one could argue that Bell was anointed with the mantle of inventor because he had an ambitious father-in-law who sued all rivals to Bell's patents and because Bell was a superb patent witness who successfully marshaled the scientific community to his defense (cf. Hounshell 1976).

But there is an aspect of the inventor question that our framework can help answer. Did Bell "borrow" or "steal" the variable resistance transmitter from Gray, as Taylor implies?

10.8.1. Elisha Gray's Speaking Telegraph

Gray was a prolific inventor who had a long string of telegraph patents, including one for a harmonic multiple telegraph. His initial mental model for a speaking telegraph was based on the lover's telegraph, which consisted of a string stretched tightly between two cones; vibrations of one diaphragm were transmitted to the other via the string. Why couldn't this mechanical motion be translated into an electrical current?

In his caveat, he included a speaking cavity similar to the cone on the lover's telegraph and extended a wire from the side of it to one of his standard mechanical representations: a magneto receiver with a resonant cavity that looked like a beaker. But where the string in the lover's telegraph was attached to the membrane directly, Gray substituted another mechanical representation. He extended a glass rod from the cone into a beaker of water. When one spoke into the cavity, the needle vibrated up and down, increasing or decreasing the amount of water between it and the contact at the bottom of the beaker. Figure 10-8 shows his first sketch of such a device, on February 11, 1876—a full month before Bell's successful liquid transmitter.

In patent testimony, Gray claimed that the "fact that the longitudinal movement (in water or a fluid of poor conducting quality) of a wire or some good conductor of electricity, with reference to another wire or metal conductor, produced variations in the resistance of an electric circuit proportional to the amplitude of movement, was old in the art at that time; so that the last link of knowledge necessary to solve the problem in my mind was furnished in the capabilities of the longitudinal vibrations of the before-mentioned so-called lover's telegraph" (*Telephone Suits*, 1880:125). He also noted that "just previous to the time I saw the 'lover's telegraph' . . . I had been experimenting with diaphragms of animal membrane in connection with a certain class of receivers; and during these experiments I used several small drums, attaching the centre of the drum-head to what I call a 'rattler', designed to operate a local circuit. In experimenting with these drum-heads, I observed that they were very sensitive to noises made in the room from any source" (*Telephone Suits*, p. 125). In other words, Gray fleshed out his mental model with mechanical representations that either he had developed or were "old in the art of the time."

On February 19, Bell was told that his proposed telephone patent of February 14 was in interference with another inventor's pending caveat (Gray's), and that Bell's application would therefore be suspended for 90

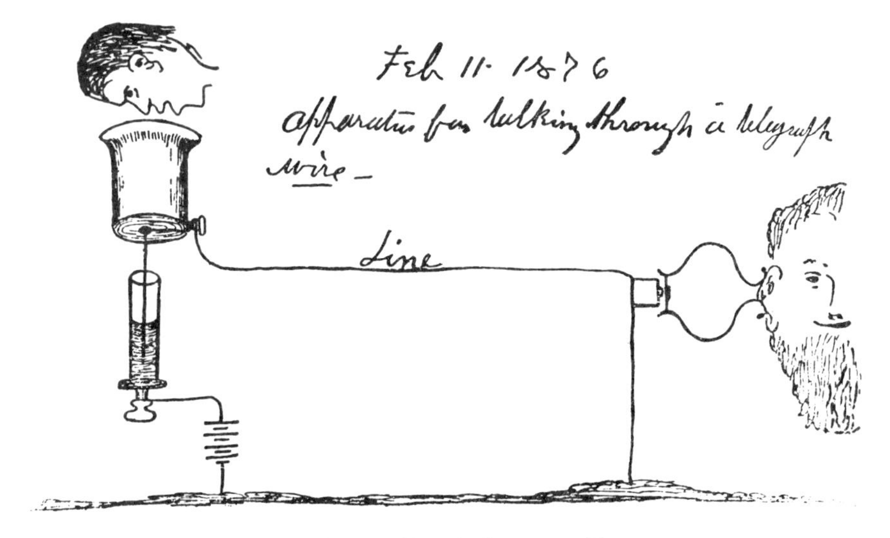

FIGURE 10-8 ELISHA GRAY'S SPEAKING TELEGRAPH
Elisha Gray's sketch of a speaking telegraph is dated February 11, 1876. When the person on the left speaks into the cylindrical cone, the membrane at the bottom vibrates, which causes the needle to vary its depth in the water, thereby changing the resistance of the circuit and transmitting the sound to the person on the right. (Drawing is from U.S. Supreme Court, 1887, *The Telephone Appeals,* Boston Alfred Mudge & Son, Law Printers, 441.)

days. But on February 25, Bell's solicitors were informed that the interference involved a misunderstanding of Bell's legal rights, and the suspension was withdrawn (Bell 1908:434).

Bell wanted to discover the source of the interference, so he went to see the patent examiner, who pointed to the line in Bell's patent that was in conflict with Gray's, a passage in which Bell spoke about vibrating a wire in mercury or other liquid included in the circuit.

Bell had actually inserted that line after submitting the patent, in response to another interference, this time with one of his own earlier patents. In this earlier, 1875 patent Bell had made references to achieving an intermittent current without making an "absolute break in the current"; in the later patent, he claimed he would make an undulating current via similar means (see Bell 1908:92). The difference was subtle, but critical, and went to the heart of the patent. The familiar Reis telephone produced an intermittent current, but not an undulating one. Bell added the line about vibrating a needle in mercury or other liquid to clarify the difference. It was this line that allowed him to establish priority over Gray in the matter of variable resistance.

So, on March 10, 1876, Bell succeeded in transmitting speech with a device that bears a superficial resemblance to Gray's. Both employed needles inserted into water. Indeed, this led Gray in 1901 to claim "that I had shown him (Bell) how to construct the telephone with which he obtained his first results" (Taylor, unpublished manuscript, IX, p. 6).

Bell, in contrast, claimed the liquid variable resistance transmitter had come from an earlier patent of his for a spark arrester, which involved inserting contacts into water. His early Helmholtz experiments had involved dipping needles in and out of dishes of mercury. It is certainly possible that Bell got the idea of using a liquid transmitter indirectly from Gray via the examiner at the patent office, but he did not have to borrow the mechanical representation.

In effect, Gray and Bell were operating from different mental models, though their liquid transmitters were superficially similar. Bell's device worked by increasing the area of the needle in contact with the water; therefore, the contact with the other end of the circuit could be off to the side and not at the bottom of the dish. Gray's device, in contrast, depended on the distance between contacts (cf. Bruce 1973).

Furthermore, Gray's patent included the possibility that a series of liquid rheostats and speaking tubes would be required to transmit speech. This multiple-transmitter mental model was derived from Gray's experience with harmonic telegraph variations; indeed, he saw the speaking telegraph as another form of harmonic multiple telegraph (Kingsbury 1915:102–106).

In conclusion, using a liquid to create variable resistance was for Bell only one of several ways in which speech could be translated into an undulating current. For example, Bell's patent covered a wide range of magneto transmitters, exemplified by the Gallows telephone. Gray's patent really covered only the liquid transmitter and therefore represents a narrower mental model based primarily on a single mechanical representation. Gray thought that the magneto devices preferred by Bell could be used only as receivers; they would never be powerful enough to transmit speech.[28] Bell's famous experiment of June 2, 1875, had shown him that a single reed relay could faithfully transmit the subtleties of the voice over a wire. The interesting question is why the examiners allowed Bell to patent a mental model.

After the initial ruling by the patent office, Gray conceded the field to Bell, in part because backers such as William Orton and Enos Barton urged him not to waste time with such an impractical invention when the multiple telegraph still beckoned (Prescott 1878). In a letter to his attorney on November 1, 1876, Gray wrote, "As to Bell's talking telegraph, it only creates interest in scientific circles, and, as a scientific toy, it is beautiful; but we can already do more with a wire in a given time than by talking, so that its commercial value will be limited so far at least as it relates to the telegraphic service" (Bruce 1973:210). Hubbard, Bell's principal backer,

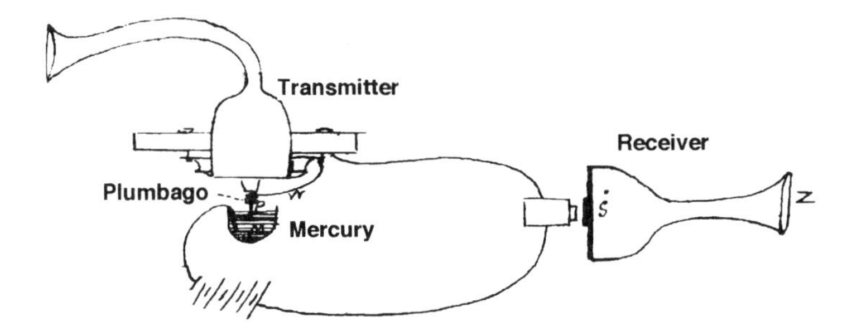

FIGURE 10-9 PLUMBAGO POINT IN MERCURY
A plumbago or pencil point on the transmitter vibrates in mercury; the resultant undulating current is transmitted to the receiver. (From "Experiments Made by A. Graham Bell" (Vol I), April 2, 1876, p. 94.)

also wanted Bell to focus on a multiple telegraph, but Bell's experience with teaching the deaf made him aware of the importance of a device that could transmit human speech.

After March 10, Bell continued on a long series of experiments based on the idea of a variable resistance transmitter (for a good account, see Finn 1966). But, in the end, he returned to his original ear mental model, abandoning the use of liquids for a metal diaphragm that moved through a magnetic field, thereby inducing an undulating current.

Indeed, Bell seems to have used this series of experiments with liquids more to confirm his understanding of undulating currents than to replace his original mental model with one based on variable resistance. The best evidence for this notion comes from his thoughts on a transmitter he tested on April 5, 1877; it involved a pencil point dipped in a mercury cup (see Figure 10-9). The pencil was so short that occasionally the speaker's voice would cause it to lift out of the mercury altogether; at this point, the undulatory current would switch to an intermittent one and a bright spark would appear. Bell thought he could hear the difference between the two types of current: "so long as the plumbago never left the mercury I could hear not only the pitch of the sound, but could recognize the quality or timbre of Mr. Richardson's voice. When the spark appeared . . . I could hear, it is true, the pitch of Mr. Richardson's voice, and that very loudly, but the *quality* had gone" (Bell 1876b:95; underlining —italics here—is Bell's).

Thus, the main result of the liquid variable resistance studies seems to have been a confirmation of the importance of an undulating current, which suggests that inventors, like scientists, might find confirmation a useful pursuit heuristic (see section 3.7).

But to say that Bell obtained a confirmatory result with these variable resistance experiments is not to say that this was his original goal. This illustrates one of the difficulties with computational simulations: a program called BELL would have to be given a goal such as "develop a telephone" as well as a set of heuristics. In contrast, Bell originally set out to invent a harmonic multiple telegraph and, even at this late stage, he was still actively considering the relative potential of intermittent and undulatory currents for such a device. Bell was unique in his appreciation of the importance of transmitting the spoken voice; to other inventors, like Gray and Edison, the real challenge was the multiple telegraph, a far more practical invention. Because Bell worked with those who could not hear, he understood how much is communicated via speech.

10.8.2. Edison and the Carbon Transmitter

"But if Bell had known anything about electricity, he never would have invented the telephone" (Moses G. Farmer, quoted in Watson 1913:14).

As noted earlier, Gray was surprised that Bell would consider using a magneto device as a transmitter. His concern was well-placed; Bell's magneto transmitters never performed very well. Bell and Watson demonstrated them in lecture halls, and the demonstration depended greatly on Watson's booming voice and Bell's skill as a master of ceremonies:

> When transmission was poor, Bell told the audience in advance what they should expect to hear; and, of course, most of them heard it. When all else failed, Watson had the knack of somehow making the audience catch one particular phrase: "Do you understand what I say?" . . . The press did not always praise the quality of transmission, on occasion likening it to "someone a mile away being smothered" or "talking with his mouth full and his head in a barrel," but it assured the reader that the sounds got more intelligible with some practice in listening. (Bruce 1973:224)

When Hubbard offered Western Union Bell's telephone patent for $100,000, they chose not to buy it: "What can we do with such an electrical toy?" President Orton is said to have remarked (Josephson 1959:141). As noted above, Orton actively discouraged Gray from pursuing his telephonic ideas.

But as Bell's fledgling company began to install telephones, Western Union became nervous and hired Thomas Edison to build an improved telephone that would compete with Bell's. At this stage of his career, Edison was still working primarily as an "inventor for hire," performing specific projects for organizations such as Western Union. He built his Menlo Park laboratory in the spring of 1876 and needed contracts like the one from Western Union to sustain it while he began projects of his own, including the electric light. Menlo Park provided Edison with a research team and his own shop, where he could turn out variations of the telephone at a far higher rate and with far greater mechanical skill than Bell.

The problem with Bell's magneto telephone was that it did not generate much volume, and it could be used only over short distances. Edison improved the volume in two ways: (1) by developing a separate transmitter; (2) by sending the current from the transmitter through the primary circuit of an induction coil. The first idea was not new; Bell's patent and Gray's caveat included the possibility that a separate variable resistance transmitter could be combined with a magneto receiver. But Edison recognized the need for a far better variable resistance transmitter than the one developed by Bell. As he said to an officer of Western Union, "You need have no alarm about Bell's monopoly as there are several things he will have to discover before [the telephone] will be practicable" (Josephson 1959:144).

W. Bernard Carlson and I, along with an undergraduate research team, are trying to reconstruct Edison's process from sketches in his notebooks and a few surviving artifacts. We can only grossly oversimplify Edison's process here. (For more details, see Gorman and Carlson 1990.)

While Bell used his research with liquid variable resistance to confirm the potential value of magneto transmitters, Edison decided variable resistance was the principle on which an effective transmitter should be based. From Edison's perspective, Bell had simply modified an earlier idea of Reis's in developing his variable resistance telephone. Reis had a needle making or breaking contact with a cup of mercury; Bell had a needle immersed in water, varying its distance from the bottom, creating a liquid rheostat.

Edison had used a liquid rheostat in a telegraph patent in 1873. In essence, the Reis telephone served as Edison's mental model; it was a mental model and not a mechanical representation because for Edison it was a source of multiple slots or variations. For example, whereas Bell had experimented with only a few different arrangements between contacts, Edison experimented with multiple points made of a variety of materials, knife edges intersecting with wires, multiple diaphragms, and dozens of different substances that could serve as variable-resistance mediums. Edison also had a far richer store of mechanical representations he could draw on to build these different experimental devices, and he worked as leader of a research team.

Edison realized that liquids would be of limited usefulness in a marketable telephone transmitter. What would happen if they dried up or leaked? Therefore, he began working with other variable resistance materials; for example, he replaced the dish of water with a film of graphite. He also experimented with alternative mechanical arrangements, experimenting with different combinations of diaphragms, switches, and resistors.

Unlike Bell, Edison was comfortable with simultaneously pursuing multiple alternatives. We suspect that Edison's strategy was more than just a matter of his having a large research team; he could, instead, have focused the entire team on a single line of inquiry until it was exhausted.

Edison's experiments focused increasingly on carbon. He decided he needed a carbon compound which was very sensitive to small changes in

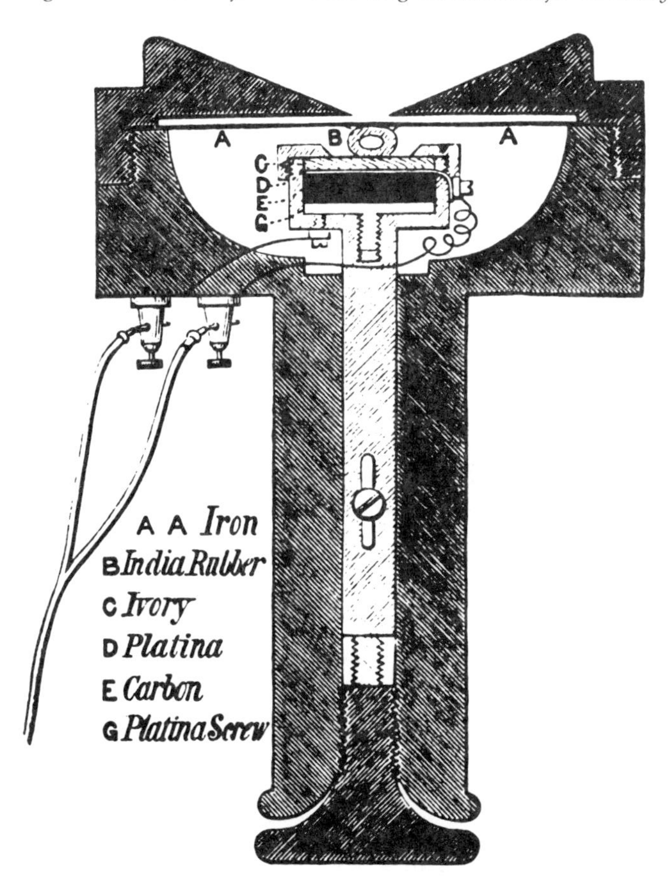

FIGURE 10-10 EDISON'S RUBBER TUBE TRANSMITTER
Edison developed this form of carbon telephone transmitter in the fall of 1877. A rubber tube B was inserted between the carbon E and the iron diaphragm A. (Drawing is from Prescott 1884:36.)

pressure. He had his assistant, Charles Batchelor, begin a "draghunt" for the appropriate material. The draghunt is a common invention heuristic; Edison used it to obtain a material for a lamp filament, Bell used it to find the best material for making phonograph records, and modern inventors are using it to search for a superconductor. It is not a random, trial-and-error search; rather, it is governed by constraints dictated by the inventor's mental model—in this case, limited to carbon compounds possessing specific properties. Lampblack carbon turned out to be the best material.

Edison initially separated the diaphragm from the carbon by means of a small rubber tube (see Figure 10-10). He was still working on his original Reis telephone mental model, in the sense that he still felt the acoustical vibrations had to be transferred by a needle or a tube through a

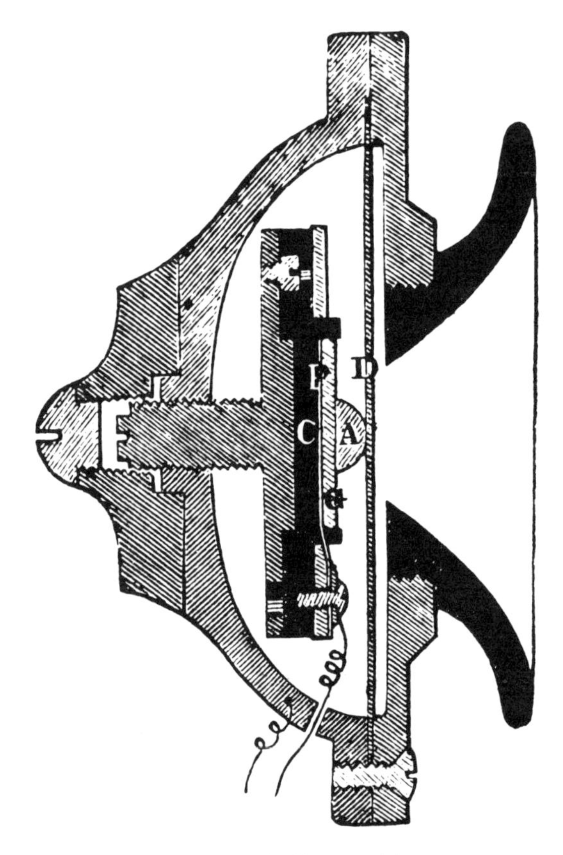

FIGURE 10-11 EDISON'S CARBON TRANSMITTER
Later version of Edison's carbon transmitter, built in the spring of 1878. The rubber tube has been replaced by an aluminum knob A, which transfers the vibrations from the iron diaphragm D to the carbon button C. (Drawing is from Prescott 1884:167.)

variable resistance medium. But eventually he put the carbon directly onto the diaphragm and obtained the best result. He had kept the idea of variable resistance but had changed his mental model of the way it worked. This new arrangement, including the induction coil, gave superior results in tests conducted between New York and Philadelphia in early 1878 (see Figure 10-11).

Edison's new carbon transmitter allowed Western Union to build its own telephone exchanges in competition with the new Bell Telephone Company. Eventually, Western Union sold its rights to Edison's carbon telephone to the Bell Corporation, in part because the latter found a patent by Emile Berliner that interfered with Edison's attempt to patent his transmitter.

10.8.3. Bell, Edison, Gray, and the Framework

Bruce (1973) compares the styles of Bell and Edison, describing the former as more of an amateur than the latter. But Bruce's account focuses more on motives for invention than on cognitive processes; he claims that Edison was more market-oriented and practical than Bell, who pursued projects like flying machines, the photophone, and a new theory of gravity. Edison also preferred large research teams; Bell worked best with a single assistant. But like all stereotypes, this one is an oversimplification; when Edison, the professional inventor, "leapfrogged" Bell on the telephone, Bell returned the favor on the phonograph, organizing a small research team and providing the financial resources.

Hounshell (1976), in comparing the styles of Bell and Gray, noted that the former identified himself more with the scientific community whereas the latter saw himself as a professional inventor. To put this in terms of investment heuristics (see 5.1.2), Bell sought deliberately to do what he could to gain acceptance from the scientific community, including gaining the blessing of Joseph Henry, Sir William Thomson, and others. Gray instead focused on popularization and commercialization of his multiple telegraph inventions. According to Hounshell, this mobilization of the scientific community was one of the keys to Bell's success.

These earlier comparisons are valuable, but they underplay the central cognitive differences between these three inventors. In terms of the framework, Bell focused more on developing an adequate mental model. It was he who recognized that speech could be translated into a continuous or "undulating" current via magnetic induction. No professional telegraph inventor would have taken such an approach seriously. But Bell's "follow the analogy of nature" heuristic and his experience with such devices as the phonautograph led him to believe that a speaking telegraph could be fashioned after the human ear.

Gray developed a mental model of the telephone based on the lover's telegraph, which showed him that "all the conditions necessary for the transmission of an articulate word were contained in any single vibrating point" and that the problem was to "reproduce electrically the same motions made mechanically at the center of the diaphragm" (*Telephone Suits*, 1880:125).

Now the effect of these motions would be to produce an undulating current, but Gray did not see the problem in these terms, and that is reflected in his caveat: his telephone worked by moving a needle up and down in water, just like the back-and-forth motion of the diaphragm on a lover's telegraph. Bell's liquid transmitter worked by alternately increasing and decreasing the surface area of the vibrating contact; he arrived at a needle only after experimenting with tuning forks and bells. When Bell was satisfied that he could produce an undulating current, he returned to his ear mental model and the magneto transmitter; Gray was surprised that such a design worked at all, and was sure that it would not produce sufficient current for effective long-distance communication.

Like Gray, Edison knew the fluctuations of a membrane had to be amplified by inserting a high-resistance medium between contacts. He based his mental model on an actual device, the Reis telephone, in part because linking his telephone to the principles embodied in Reis's device gave him a potential way around Bell's powerful patent. If Edison could build a successful transmitter based on the Reis device, that would undermine Bell's claim to originality. Here investment heuristics may play a role in the formation of a mental model.

Whereas Bell's greatest strength lay in his understanding of speech, Edison's greatest strength lay in the craft or "hands-on" knowledge possessed by him and his assistants. On the telephone, Bell was more of a "top-down" inventor, conducting thought experiments to clarify his mental models and eventually patenting an idea he had not even "reduced to practice." Edison was more of a "bottom-up" inventor, in that he was assigned the task by Western Union, decided that an existing device was the best place to start, and generated hundreds of mechanical variations which give us clues to his mental models. Gray comes closer to Edison in that he also worked more "bottom up" from the bench than "top down" from mental models.

Perhaps putting this distinction in other terms will make it clearer. Klahr and Dunbar (1988) discussed how subjects discovering the function of a repeat key on a Big Trak could be divided into theorists and experimenters based on their problem-solving styles (see section 9.2.2). Similarly, we might argue that Bell had a more "theoretical" style than either Edison or Gray. As Bell said: "It became evident to me, that with my own rude workmanship, and with the limited time and means at my disposal, I could not hope to construct any better models. I therefore from this time (November, 1873) devoted less time to practical experiment than to the theoretical development of the details of the invention" (Bell 1876a:8). At this point, the cognitive and the social intersect; Bell may have employed such a style in part because it was more likely to win acceptance from the scientific community (cf. Hounshell 1976).

As we have seen, we can identify specific differences among the mental models of these three inventors. We can also compare their mechanical representations. The sorts of components or "standard solutions" preferred by Gray begin to bear a closer resemblance to Bell's: for example, both worked extensively with variations on musical instruments, such as tuning forks, pianos, and violin strings. But Gray developed a larger set of transmitters and receivers he could use as mechanical representations. Edison, in contrast, had a rich store of mechanical representations based on carbon that neither Bell nor Gray had much experience with.

In terms of heuristics, there are some immediate similarities. Both Bell and Edison employed the "draghunt" while searching for the best variable resistance medium. However, Edison's was far more sophisticated, involving hundreds of different materials. Bell liked to employ a "holding constant" heuristic, adding or subtracting one or two mechanical representations at a time to note their effect on transmission. Edison, in contrast, used what Bruner, Goodnow, and Austin (1956) call a "focused gambling" heuristic, in which one varies multiple attributes simultaneously in hopes of coming up with a successful device or theory. Gray employed a heuristic I call "patent all combinations": he would try virtually every possible transmitter-receiver pair and patent all combinations that yielded promising results. In fact, about a year after his telephone caveat was dismissed, he tried to revive it by pairing his liquid transmitter with other receivers he had designed.[29]

10.8.4. Confirmation, Disconfirmation, and Invention

Inventors also employ confirmation and disconfirmation, especially the "confirm early, disconfirm late" heuristic: in the early stages, an inventor typically looks for devices or variations that yield promising results; later, he or she will test the "winners" under conditions approximating their actual use. As in science, these efforts are often done by teams. Consider the Sidewinder (Westrum and Wilcox 1989). This air-to-air missile was developed by a team lead by Bill McLean. They had to build a device that would outcompete alternatives favored by the military bureaucracy. More than that, they had to constantly prevent the project from being shut down or "disconfirmed" by visitors from the "top brass." They developed a clever set of demonstrations of the heat-seeking apparatus, mounting it so it could point a radar antenna at a distant aircraft. The point is, confirmation and disconfirmation are often social processes; an inventor has to convince others that a design or device is worth pursuing.

Bell and Watson's lecture demonstrations performed this role very well. One could argue that Bell exhibited a certain amount of confirmation bias, in that he never lost faith in his magneto design even though it produced a telephone that was very difficult to hear. Similarly, in section 8.2 we discussed how Edison used only the first part of the "confirm early, disconfirm late" heuristic in his pursuit of an "etheric force." In his pursuit of the telephone, however, Edison was far more exacting, building and abandoning dozens of designs that did not faithfully reproduce a full range of speech sounds. Had he been as exacting with the phonograph, Bell would never have "leapfrogged" him.

10.9. THE FUTURE OF A FRAMEWORK

These remarks about Bell, Gray, and Edison are preliminary and speculative, but they illustrate the potential strengths and weaknesses of the cognitive framework that serves as the focus of this chapter. The framework itself is a kind of mental model—incomplete, unstable, constantly evolving. It alerts us to look for three major components of an inventor's mental process: the working, three-dimensional "mental picture" of a new technology; the strategies or procedures one uses to complete and/or alter that picture; and the specific mechanical parts or "tricks" one borrows from other inventions. If the framework is useful it will guide us to more specific and detailed comparisons between the mental processes of inventors and scientists and will allow us to talk about "style" in more rigorous terms.

The danger is that the framework will merely introduce a new set of terms which we can use to say the same kinds of things that have been said in the past: old wine in new bottles. The comparison above between Edison and Bell raises this possibility; to say that the former is experimental and the latter is theoretical is to make only a general statement, strengthened by linking it to experimental studies.

But the framework does allow us to be more specific about these stylistic differences. We can show how two sketches by Bell and Gray, although superficially similar, in fact reflect different mental models. Whereas Bell struggled to articulate theoretical ideas about how a telephone ought to work, Edison appears to be trying a million-and-one random variations. But the variations are not random; we have shown how they can be organized under slots in his mental model based on the Reis telephone (see Carlson and Gorman 1990).

One of the new sociologists of scientific knowledge might criticize this whole approach on the grounds that it falls into a mentalistic muddle. Why assume that Bell's ear analogy is in his "mind"? Why not simply treat it as an inscription? One could construct maps on this basis, drawing slots on inscriptions and organizing actual devices into flowcharts.

At a recent convention of the Society for the Social Studies of Science, I responded to this attack on mentalism by pointing out that I had a mind, and that I could imagine things that I never "inscribed," or directly discussed with anyone else. But the critic could have countered by arguing that one cannot study what is not inscribed. This point resembles John Watson's and B. F. Skinner's classic argument that to be scientific, psychology must deal only with observable behavior.[30]

Behaviorism failed as a movement in part because to understand how human beings solve problems one has to know how they represent them (see section 3.9). Similarly, to say that mental events are simply the byproducts of social interaction misses the way in which minds transform knowledge. Consider Bell's mental model of the telephone. Certainly, one can show how parts of it were derived from interaction with others: for example, by working with Clarence Blake on the phonautograph and by studying Helmholtz. But Bell transformed these experiences into something unique, something no one else would have considered. Indeed, other telegraph inventors thought Bell's idea bordered on the ridiculous.

Furthermore, Bell, Gray, and Edison could imagine devices they never built or sketched. A slot diagram can give us clues to alternatives that never appear as inscriptions or devices. For example, Bell was not limited to the two alternatives sketched in Figure 10-1; indeed, his next line of experiments shows that he could as easily imagine a tuning fork and an electromagnet inserted in place of the ossicles, though he never produced a sketch illustrating that idea.

To assert that inventors have minds is not to make them individual cognizers operating in a social vacuum. That is one of the problems with current computer simulations of science. This cognitive framework forces one to "follow inventors around," plunging into the fine details of the invention process, including the social matrix in which the inventor operates. How did Bell transform Helmholtz? What did Gray's backers say about the prospects for a telephone? What constraints did Western Union set for Edison? What does it mean to consider oneself an "inventor" at the end of the nineteenth century in America? As noted earlier, mechanical representations and heuristics can be divided into those held in common and those unique to a particular individual. But even an inventor's unique heuristics and mechanical representations are at least partly derived from other sources.

Why was the telephone so successful? It was not simply "a good idea whose time had come"; indeed, multiple telegraphy was the obvious technological advance. The framework does not specifically address the complex social negotiations required to create a need for an invention, or the process by which one person is labeled *the* inventor. Bell did build the first working device, but he used a principle known also to Gray, and without Edison's transmitter, Bell's telephone never would have performed satisfactorily. One could adapt the framework to discuss an inventor's mental model of the potential user for a device, and also his or her marketing heuristics. Carlson and Gorman (1990) have sketched how Edison's marketing ideas were integral to his mental model for a device to present moving pictures.

10.10. SIMULATING INVENTION

This chapter has, I hope, shown how a cognitive framework for studying invention can emerge out of experimental simulations, though the framework is continually being revised as it confronts the historical "data." However, the framework, and the study of telephone inventors, ought to suggest some ideas for future simulation.

Consider what would happen if the 2-4-6 task were supplemented by a task simulating Bell's invention process. Students could be given one of

Bell's standard devices, divided into slots, in lieu of a starting 2-4-6 triple. They could also be given a set of Bell's mechanical representations, including descriptions of how they operated. The goal would be to discover which assemblage of mechanical representations would best reproduce a tone; the students could experiment by combining icons of the mechanical representations on a computer and listening to the results. (This sort of visual task with auditory feedback could be implemented in hypercard on a Macintosh computer.) Obviously, similar tasks could be based on other inventors.

What variables might one investigate with such a simulation? As a start, one could explore the role of confirmation and disconfirmation in the invention process by asking subjects to test a device. One could, for example, begin with one of Bell's magneto telephones and ask subjects to determine whether it is an effective design. Would they adopt a confirmatory strategy and focus on improving the existing device, or would they try to disconfirm it by testing alternative arrangements? One could instruct subjects to confirm, or disconfirm, or to confirm early, disconfirm late, supplementing these instructions with examples.

One could also instruct them to employ a number of the heuristics used by the telephone inventors. For example, in addition to stating hypotheses about the function of their devices, subjects could be encouraged to develop mental models as they progress and produce slot diagrams of them. To facilitate the development of useful mental models, one could give subjects examples like Bell's ear analogy. One could also explore the role of slot diagrams independently of mental models; subjects could be encouraged to draw new slots on experimental devices they created in the course of their experiments.

Following Klahr and Dunbar (see section 9.2.2), one could also investigate whether subjects evolved different invention styles. Would some, for example, prefer to work in a kind of hypothesis or mental model space, doing thought experiments to explore potential variations before constructing tests? Would others work more in a kind of experiment space, building and testing almost all variations on the computer screen?

Naturally, an experimental simulation of the invention of the telephone would require that subjects be given a great deal of background information. Therefore, the experiment could also have educational value: students could learn the history of a major invention by working with the inventor's mechanical representations. This interactive learning environment would teach more than history; it could help students understand and appreciate technological innovation, and even help them explore principles of good design. I have always designed my experiments to have maximum educational value for the participants. Why not combine teaching and research by having students learn as they provide data?

Obviously, simulations could be developed for other major inventions as well, and could even incorporate features like the competition between

Gray and Bell. But one could also develop original invention tasks that, like the 2-4-6 task, were domain-independent and relatively context-free. One could construct a variety of "micro-worlds" in which subjects had to build structures using different materials and connections, or design devices like the Big Trak and test them under varying terrain conditions. The possibilities are limited only by the imagination.

In conclusion, frameworks developed in ecologically valid case studies need to be complemented by externally valid experimental simulations. From frameworks and cases we can develop ideas that need further testing under ideal, artificial conditions: ideas about the role of specific heuristics, about the advantages and disadvantages of different forms of representations, about principles of good design. But laboratory simulations, in turn, need to be complemented by case studies of other inventions, e.g., a twentieth-century effort involving international research teams.

The canals on Mars case discussed in 8.3.2 shows more clearly how simulation can complement case study. Indeed, the framework outlined in this chapter could be applied to Lowell and other scientists involved in the controversy. What were the origins of Lowell's mental model of Mars? How did it evolve in response to evidence and critics? What observation heuristics did Lowell use? How did new mechanical representations developed at his observatory and others affect the resolution of the controversy?

10.10.1. Computational Simulation of Invention

How might computer simulations complement the comparison of telephone inventors outlined in the last chapter? One could certainly imagine trying to build BELL or GRAY programs, which would use heuristics on a variety of levels to "construct" devices from sets of propositions designed to represent different functions. A tuning fork, for example, could be translated into a list of attributes that the program could put in a slot on an experimental apparatus it was "building."

One could go a step farther and attach visual images to these lists of propositions, so that the program would actually manipulate images on a screen: one could see it try to invent. Results of all tests would probably have to be supplied by the programmer, though it might be possible to set up a separate program that would match the computer's experiments with the sort of results obtained by Bell, and give the BELL program feedback.

The advantage of such a simulation is that it might provide additional conceptual rigor, forcing one to define aspects of the framework in ways that could be implemented on a computer. This would help prevent the framework from becoming so idiosyncratic that it cannot be used by others on new cases. Computer simulation can be a useful heuristic for forcing one to be specific about terms like *mental model* and *mechanical representation*.

The disadvantage is the Procrustean bed problem. Concepts like *mental model* may not be implementable in any current computer architecture. Does this mean they ought to be abandoned? I think not.

Instead, I would prefer to develop a program that would operate as an expert assistant, helping historians and psychologists explore the invention process. Preliminary steps in this direction are being taken by Thomas Addis and David Gooding, who are developing functional, computational representations of aspects of Faraday's processes (cf. Gooding and Addis 1990). Instead of developing programs designed to demonstrate that machines can invent, this approach emphasizes developing programs to help human beings understand invention.

How might such a program work on Bell? Well, for a start, it could help clarify what we mean when we say he "opens a slot"—by taking every instance where we claim Bell has opened "a slot" noting the inputs and outputs, and then predicting where an "open slot" ought to occur on a different section of the map. Obviously, one could do this operation across as well as within inventors. We could compare the program's predictions with our own intuitions to get a better understanding of what we mean by "opening a slot."

We hope that our framework, in combination with the idea of creating flowchart maps of the invention process based on slot diagrams,[31] will provide a methodology that can be applied to comparisons between other inventors or scientists. An expert assistant program could be used as part of a package designed to add rigor to this framework and help others use it on novel comparisons.

10.11. TRANSCENDING TRADITIONAL DICHOTOMIES

If there is a moral to this lengthy autobiography of a research program, it is that to understand technoscientific thinking one must transcend traditional disciplinary and methodological ideologies. As I write this, the Middle East is engulfed in war, as are Cambodia and many other parts of the world. The products of technoscience play an important role in these conflicts, and also in the rising tide of democracy sweeping the world. We must understand the bombing of Hiroshima as well as the birth of relativity; no discipline or methodology or perspective can accomplish this goal alone.

What is alarming is the way different methodologies get divided into separate discourse communities, often competing for resources and respect. Clearly, simulation and observation, model and framework have to be closely linked. These are not either/or choices; they are complementary ways of knowing.

This does not mean that every scholar must be an experimental psychologist, a historian, and a sociologist; it does mean that those pursuing one method or approach must be willing to "open their black boxes" to others representing different perspectives, and even occasionally to collaborate. To study invention, I needed to evolve a partnership with a historian of technology. We have had to learn to synthesize perspectives and approaches that others would have viewed as contradictory.

I opened this book by quoting Latour's proposal for a moratorium on cognitive studies. (Contra his own moratorium, Latour has shown great sympathy for, and given us valuable suggestions on, our telephone research.) Let me close by proposing a moratorium on moratoria, with one caveat: "let a hundred flowers bloom" in terms of different approaches, but let there be open communication among the variety of new flora and fauna—communication marked by a lack of contempt or arrogance. Technoscience studies will only succeed when those engaged in following scientists and inventors around work closely with those involved in simulating their activities.

The fragmentation between disciplines and perspectives has its worst effect in the classroom. I remember during the heyday of behaviorism it was possible to take psychology classes in which alternative views were either ridiculed or not covered at all. To help students evaluate hypotheses like Von Daniken's (see Introduction), it is not sufficient to teach them how to falsify, though that may help. Nor can we make sure that they have knowledge of all the relevant areas of science and technology. Instead, we need more opportunities for interdisciplinary collaboration in the classroom, encouraging students to plunge into the thicket of issues surrounding technoscience. If nothing else, they will become better global citizens, able to evaluate the promise and perils of technoscience in a time of smart weapons, supercomputers, and greenhouse gases. These problems and opportunities are not the exclusive prerogative of any strong program; they can only be solved if we learn to collaborate on a global scale.

Notes

1. FALSIFICATION IN THE LABORATORY

1. Griggs and Cox explained their results in terms of memory-cuing: "performance on the selection task is significantly facilitated when the presentation environment of the task permits the subject to recall past experience with the content of the problem, the relationship (rule) expressed, and a counterexample to the rule" (Griggs 1983:24). A variety of other explanations besides verification bias and memory-cuing have been offered for results on the selection task, including psycholinguistic matching biases, illicit conversions from conditional to biconditional rules, and covert changes in the task caused by subtle shifts in instructions, but a discussion of these lies beyond the scope of this book (see Evans 1983 for a series of papers that review this research and Cosmides 1989 for a new hypothesis: that results on the selection task can be explained in terms of social exchange).

2. There has been a long philosophical debate over whether experimental results from tasks like the selection task provide evidence that human beings are irrational (cf. Stich 1985) or merely "trick" people in a way that makes them appear to respond irrationally (cf. Cohen 1981). Tweney and Doherty (1983) argue that psychological studies of scientific reasoning are relevant to epistemology but do not conclusively demonstrate that scientists are or are not rational. In chapters 7 and 8, we will have more to say about the role of experiments in making normative judgments about science.

2. FALSIFICATION AND THE SEARCH FOR TRUTH

3. To be statistically significant, a result must have a probability of less than one in twenty of having occurred by chance. When multiple dependent measures are used, I have avoided "probability pyramiding" by using multivariate tests. In this book I have chosen to omit the specifics of the statistical tests used, on the grounds that they will be of interest only to a specialized audience. Details can be found in the versions of these experiments that appear in scientific journals; citations are provided throughout the book.

4. Professor Evans was a reviewer and editor on many of the papers I subsequently submitted to British journals, and I found him a fair but demanding critic who made suggestions that strengthened several of my pieces.

5. I would like to propose an allied metaphor—the research factory, exemplified by certain institutes and laboratories that become centers for promoting a certain way of doing research. The products, in this case, are students, publications, and grants. The work of Simon, Anderson, and others at Carnegie Mellon serves as an example in cognitive science: they have been successful at promoting a particular kind of computational view of problem solving (see chapters 9 and 10), in part because of their own success at attracting funding and exporting students. A research factory, with other spin-off "plants," may form the driving core of an invisible college. To push this analogy, my own research program initially resembled a small cottage industry, partly spun off from the Bowling Green factory but partly the result of independent development.

3. WHEN FALSIFICATION FAILS

6. Whether an explanation involving mental representations is consistent or inconsistent with one involving logical competencies is an issue we will leave for

the philosophers to settle (see note 2). Clearly, mental representations play an important role in subjects' and scientists' choice of evidence on the selection task; when the appropriate mental representation is invoked, the importance of potentially falsificatory data is "obvious." We will have a bit more to say about this in chapter 10 when we discuss P. N. Johnson-Laird's (1983) research on the role of mental models in solving syllogistic reasoning.

7. For example, I recall a specialist in rhetoric confidently telling his audience that Kuhn had shown that science does not progress. Had Kuhn been in the audience, I fear he would have done violence to the speaker! Kuhn himself recalls a similar incident in his "Reflections on my critics" (1970:263): "During a meeting I was talking to a usually far distant friend and colleague whom I knew, from a published review, to be enthusiastic about my book. She turned to me and said, 'Well, Tom, it seems to me that your biggest problem now is showing in what sense science can be empirical.' My jaw dropped and still sags slightly. I have total visual recall of that scene and of no other since de Gaulle's entry into Paris in 1944."

Kuhn attributes much of the misunderstanding to the novelty of his perspective, but he is also responsible for introducing a new set of terms that are often used in baffling and inconsistent ways (see Masterman 1970).

8. To support his model of the growth of science, Lakatos transforms historical cases into rational reconstructions: "In writing a historical case study, one should, I think, adopt the following procedure: (1) one gives a rational reconstruction; (2) one tries to compare this rational reconstruction with actual history and to criticize both one's rational reconstruction for lack of historicity and the actual history for lack of rationality. Thus any historical study must be preceded by a heuristic study: history of science without philosophy of science is blind" (1978:52–53).

Kuhn (1970) points out that Lakatos's rational reconstructions are full of footnotes explaining their inaccuracies; at one point, Lakatos (1978) warns the reader that, "Some statements are to be taken not with a grain, but with tons, of salt" (p. 55). Kuhn claims that these footnotes illustrate the difference between the way philosophers and historians approach science: "The problem is not that philosophers are likely to make errors—Lakatos knows the facts better than many historians who have written on these subjects, and historians do make egregious errors. But a historian would not include in his narrative a factual report which he knew to be false" (1970:256).

9. Paul Feyerabend attacks the prescriptive aspects of the philosophies of Popper, Kuhn, and Lakatos. His criticisms of Popper and Lakatos focus especially on Popper's idea that concepts and theories, once disseminated, exist in a kind of separate "World Three." The "First World," according to Popper, is the material; the second is consciousness; the third is made up of propositions and standards of truth—it is the world of objective knowledge, where psychological and sociological factors do not play a role. Feyerabend uses Kuhn's idea that different paradigms are incommensurable to refute the existence of World Three. By "incommensurable" Kuhn means that different paradigms don't even use the same language; one literally cannot be translated into another, and the followers of one do not even understand the other, unless they go through the dramatic "gestalt-switch" discussed earlier. So there is no objective, neutral language of ideas and therefore no World Three independent of psychological and linguistic differences between human beings. Therefore, even Lakatos's modifications of Popper cannot save rationality.

Feyerabend is not certain whether Kuhn's philosophy is primarily descriptive or prescriptive. Are his statements about normal science and revolution merely a summary of how science develops, or are they also intended as a blueprint for how science ought to grow? According to Feyerabend (1970), Kuhn seems to suggest that if fields like many of the social sciences want to make the transition from a preparadigmatic state to normal science, they ought to restrict criticism and accept one theory as a basis. But Kuhn (1970) is careful to say that the only sure way to achieve normal science is through slow maturation and that paradigms operate in disciplinary communities, not necessarily across entire fields.

Whether Kuhn is descriptive or prescriptive, Feyerabend thinks he is right about the existence of both normal and revolutionary science, but wrong to think science progresses from one to the other. In fact, he credits Lakatos with the realization that both normal and revolutionary science exist at the same time—some scientists are absorbed with solving puzzles and others are challenging fundamental ideas and developing new theories. There is value to both tenacity and theory proliferation:

> Proliferation means that there is no need to suppress even the most outlandish product of the human brain. Everyone may follow his inclinations and science, conceived as a critical enterprise, will profit from such an activity. Tenacity: this means that one is encouraged not just to follow one's inclinations, but to develop them further, to raise them, with the help of criticism (which involves a comparison with the existing alternatives) to a higher level of articulation and thereby to raise their defence to a higher level of consciousness. The interplay between proliferation and tenacity also amounts to the continuation, on a new level, of the biological development of the species and it may even increase the tendency for useful biological mutations. It may be the only possible means of preventing our species from stagnation. (Feyerabend 1970:210)

Tenacity and proliferation resemble verification and falsification, respectively. Both are necessary components of science.

4. HOW THE POSSIBILITY OF ERROR AFFECTS FALSIFICATION

10. Giora Hon argues that scientists commonly distinguish between two types of error: systematic and random.

> Systematic errors . . . obstruct the measurement from reaching the intended actual value by shifting the result—either positively or negatively—by a magnitude which may be constant or may vary in some regular fashion. By contrast, random errors are disordered in their incidence and vary accidentally in their magnitude. A simple example of the former is an incorrect calibration of the measuring instrument; undue mechanical vibrations of the equipment—vibrations which interfere with the measurement—constitute a cause of the latter. (1989:474)

Kern's system-failure error resembles Hon's random error; her measurement error resembles his systematic error. Perhaps the major difference is that system-failure error represents an extreme case of random error, where by chance or device breakdown a result that would confirm a theory appears to disconfirm it, and vice versa.

Hon goes on to propose a far broader typology of errors, covering the whole experimental process from initial conception to final conclusions. His examples include some of the cases we will discuss in subsequent chapters, e.g., Millikan, Blondlot, and Lowell.

5. PURSUING THE POSSIBILITY OF ERROR

11. I wanted to conduct a thorough replication-plus-extension of Tukey's work, using his terms in a number of different studies, so I wrote him to ask for a copy of his instructions. Experimental psychologists ought to be willing to share their instructions to subjects, because nuances in these instructions often greatly affect results: witness the difference in experimenter feedback between my studies of disconfirmation and Tweney's, a difference which only emerged when he went through my procedure.

Unfortunately, Tukey never replied. I tried to use his terms as they appeared in his published article. I was particularly interested in whether his terms would show different results when subjects were run without feedback on whether each of their guesses was right or not.

A student of mine, Judy Haarala, ran twelve subjects in each of two conditions: (1) where they received feedback on whether each guess was right or not; (2) where they received no such feedback. Subjects within each condition were split randomly between two rules: (1) ascending numbers, or (2) odd-even. We were mainly concerned whether different patterns would emerge in their use of terms. In particular, would subjects who had to test their hypotheses without experimenter feedback use terms like *disconfirm* and *eliminate* more than other subjects? The answer, in brief was no—because each individual subject seemed to use the terms in idiosyncratic ways. Some would focus on three or four terms, and use them exclusively; others would try to use all seven. Favorites were *search* and *explore*—terms that did not tell us a great deal about what subjects were really doing.

I concluded that the terms were not very helpful. Indeed, they run counter to Tukey's desire to "learn from subjects which methods or strategies are easily and/ or beneficially explored," in that they force subjects to categorize their behavior in ways that may distort their actual processes. (See also section 5.4.2.)

6. SIMULATING ACTUAL ERROR

12. One of the differences between Millikan and Ehrenhaft is that the latter tried to account for every piece of data, whereas the former realized that some of the data were erroneous and ought to be discarded: "Indeed, Millikan would have warned Ehrenhaft that using *all* readings equally, just as they come in, would be defensible only in a completely routinized situation where the chances for artifacts entering the 'open window' have become negligible" (Holton 1978:69). In other words, Ehrenhaft treated the situation as something akin to early versions of the 2-4-6 task, where the possibility of error was negligible; Millikan was acutely aware of the likelihood of error in a novel, experimental situation.

Another aspect of the controversy concerns whether Stokes's Law was applicable to irregular metal particles of the sort used by Ehrenhaft. Millikan thought it wasn't; Ehrenhaft thought it was. Events subsequently proved Millikan right (Hon 1989:485). This sort of controversy over laws governing procedures and measurements is not simulated in Eleusis or in the 2-4-6 task, but could be built into future simulations.

7. A TALE OF TWO JOURNALS

13. According to Manier, "Darwin's rhetorical strategy is basically ironic . . . it involves constructing an appearance of conformity to an ideology which he understood himself to oppose, even to the point of causing 'the fabric to totter and fall' " (Manier 1987:590). The ideology, in this case, is the sort of scientific

methodology espoused by Herschel, among others: "Darwin's rhetorical use of Herschel's methodology . . . not only did not conform to the chronology of his own scientific investigation, it was called into question by the details of that chronology" (p. 593).

Lennox (1989) argues that Darwin's thought experiments, recorded in his notebooks, initially helped him develop his theory, but in the *Origin* Darwin transformed them into a powerful persuasive tool. Lennox is more concerned with the philosophical ramifications of this use of thought experiments, so he does not link them to Manier's more sociocultural analysis.

14. At times, the language used by behavioristic referees was quite pejorative. For example, Garcia cites one referee who "informed the editor that one of our recent manuscripts would not have been acceptable even as a term paper in his or her learning class. (Unfortunately, since the review was anonymous, I was unable to properly congratulate the consultant on his or her high academic standards)" (1981:149). Spencer, Hartnett, and Mahoney (1986) found that experiences like Garcia's were not unique: 40 percent of a sample of referees' reports they analyzed contained at least 25 percent emotional rhetoric, much of which was pejorative in the case of rejections. Encouraging referees to sign their reports might eliminate much of this pejorative rhetoric. Both of *QJEP*'s referees on my manuscript signed their reports; however, none of *JEP:LMC*'s did, despite an editorial policy that explicitly encourages them to do so.

15. Mahoney's and Peters and Ceci's work suggests another heuristic: do not "study the powerful," to adopt Joan Sieber's (1989) term. She cites the case of William Epstein (1990), who did one of these pseudo-manuscript studies on social work journals; while his findings suggested evidence of confirmation bias, they were not conclusive. But his qualitative analysis revealed that "Referee reviews from prestigious as well as nonprestigious social work journals were not knowledgeable, scientifically astute or objective" (Epstein 1990:24–25).

His work was attacked as unethical by editors of several major social work journals and he had to publish his study in a journal outside the field. Similarly, reactions to Mahoney's and Ceci and Peters's studies were negative. None of these authors obtained informed consent from journal editors and reviewers in advance; perhaps if one obtained their agreement, this sort of research could be continued under less controversial circumstances. Scientific journals should promote studies of the effectiveness of their review procedures.

16. The recent literature in cognitive psychology includes a number of authors who explicitly state their intention to follow the hypothetico-deductive format. For example, Cosmides (1989) outlines a Darwinian approach to cognition and applies it to the selection task in a series of studies designed to demonstrate its superiority over rival hypotheses. Massaro (1989) employs strong inference to eliminate alternate hypotheses in a chain of studies directed at refuting the modularity thesis. We cannot know whether both of these authors actually followed the hypothetico-deductive format throughout their actual research, including pilot studies—like most researchers, they may have left out at least some of the important twists and turns in their published reports. But it is clear that the hypothetico-deductive format is publicly favored by many scientific psychologists as a way of doing research, not just reporting it. Whether following this norm really facilitates ongoing research is a matter for further empirical study; at the very least, Cosmides and Massaro indicate how this format can be used to create "a juggernaut of persuasion" (cf. Bazerman 1988).

17. Tweney, in conversation, reported having divided students in one of his classes into groups to work on his most complicated artificial universe task. The exercise apparently produced interesting results, and supports the educational value of a large-scale simulation of scientific problem solving.

8. FROM LABORATORY TO LIFE

18. A schema is a type of mental representation that is closely related to the concept of a frame, introduced by Marvin Minsky (1985). A frame is composed of slots and bins that contain information associated with an object, e.g., the frame for *bird* might include such information as "is an animal," "has feathers," "flies." Note that "flies" is not always true; this would represent a default value in a frame (Thagard 1988). Frames are precise enough to be implemented in an expert system.

Schema is a fuzzier concept which evolved, in part, to deal with the fact that concepts stored in memory are more than the lists of defining attributes required by a frame; they are also images (Gardner 1985). Schemata produce expectations that are derived from experiences with groups of instances. We expect dogs to bark and birds to fly; these expectations can be included nicely in a frame representation. But we can also imagine a typical dog or bird or chair and this prototype influences our expectations about what we will experience when we encounter a new dog or bird or chair.

9. USING TECHNOLOGY TO STUDY TECHNOSCIENCE

19. Ray (1987) illustrates the dangers of these computer models when he describes how the famous "Stonehenge is an astronomical observatory" hypothesis depended crucially on a computer simulation. Hawkins, the author of the theory, described his role as that of a "middleman" in bringing the machine's discoveries to the public eye (see Hawkins and White 1965). Later work has shown that Hawkins's thesis is in many respects incorrect. Computational models depend crucially on the assumptions of those constructing them: "garbage in, garbage out."

20. PI may also provide a more flexible response to disconfirmatory information. KEKADA treats disconfirmatory information as a surprise, and generates new hypotheses upon encountering it. In contrast, in PI a disconfirmatory result may affect only one of the subordinate rules connected to a hypothesis (see section 4.2). A good example is provided by Pinch's (1985) account of how a theorist revised his calculations of solar neutrino flux to account for an apparently disconfirmatory result. In PI, these calculations and assumptions would be modified in the light of new evidence; however, KEKADA, encountering this surprising result, would presumably have generated a whole new set of hypotheses.

21. Note that this analogy relies on surface attributes, such as color, and therefore in Gentner's (1980) terms is a weak analogy (see note 24 below). Analogy is an important mechanism in achieving Thagard's (1989) explanatory coherence as well, but he merely says that an analogy must be "explanatory," "with two analogous propositions occurring in explanations of two other propositions that are analogous to one another" (p. 437). He makes no distinction between good and bad analogies, and therefore it is not clear whether this "red" analogy would contribute to coherence. Even computational theories can be vague. Thagard's current research focuses on developing more precise computational models of analogy retrieval (Thagard, Holyoak, Nelson and Gochfeld 1990).

22. Actually, Thagard is sympathetic to the idea of a kind of computational "face-off" between competing philosophical approaches to science. Thagard (1988) argues that parallel processing systems will be necessary for this sort of comparison. Each of the multiple processors in a parallel processing system is supposed to correspond to an individual scientist, with a "controlling program," corresponding to the entire system, by which grants are awarded, articles are refereed, etc:

> The central executive would serve to collect and communicate information from the separate researchers. In actual science, professional journals serve much of this function, with journal editors and referees functioning to screen the results of research for what is worth looking at by other researchers. Individual scientists pursue their research through processes of problem solving, hypothesis formation, and experientation, but share the results of their investigation through the executive. (Thagard 1988:186–187)

Thagard cites the Connection Machine, developed by W. Daniel Hillis and others at Thinking Machines Corporation, as one of his primary examples of a parallel processing system. While Thagard does not describe this system, it is important to understand how it works before evaluating its potential as a simulator of group rationality. Basically, its current applications involve rapid comparisons of a given case with a large database, to determine the best match. For example, the Connection Machine does medical diagnosis by comparing each new patient with a large database of existing patients. If it finds no similarities, the system indicates that the new case does not fit any current diagnostic categories. If it finds a cluster of patients with the same diagnosis and that diagnosis resembles the new case, it concludes that the new case has that same diagnosis. If the cluster of similar cases includes several diagnostic categories, then the system can propose tests—e.g., taking a throat culture—to discriminate among them, and use the results of the tests to arrive at a final diagnosis. Note that this system uses no heuristics or rules; it calculates the metric used in computing similarity "on the fly" (Waltz 1987). Therefore, its controlling program is extremely simple.

Thagard's view implies that each processor in a machine of this sort will resemble an individual scientist, engaging in "problem-solving, hypothesis formation, and experimentation." But it is not clear how each of the thousands of small processors in a Connection Machine could simulate all of these higher-order functions when even a much larger computer cannot. Each of these processors is not intelligent; intelligence emerges from the whole.

Thagard also implies that the "central executive" will be extremely sophisticated, simulating the complicated political negotiations that are involved in refereeing articles and awarding grants. But the serial control program on the front-end computer relies on a few simple algorithms to carry out operations simultaneously on all of the Connection Machine's processors, calculating matches and calling for further tests. Ease and simplicity of programming is one of the machine's great advantages. Thagard's analogy between science and the Connection Machine is more confusing than illuminating.

23. Such a program might be implemented on a Connection Machine (see note 22, above). According to David Waltz (1988), to emulate human intelligence, these machines would have to possess a number of connections roughly equivalent to those in the human brain, which means that no current computer approaches the capacity to emulate human discovery processes. Waltz believes that a computer of this sort ought to be available as early as 2012. Whether such a computer could truly emulate human intelligence is still an open question.

10. A COGNITIVE FRAMEWORK FOR UNDERSTANDING TECHNOSCIENTIFIC CREATIVITY

24. And I cherish more than anything else the Analogies, my most trustworthy masters. They know all the secrets of Nature, and they ought to be least neglected in Geometry.

(Kepler, quoted in Gentner 1980:2)

There is an extensive literature on the role of analogies and metaphors in scientific thinking (see, for example, Black 1962; Hesse 1966; Edge 1974; Hoffman 1980; and Leary 1990). A discussion of this literature lies far beyond the scope of this book, but it is important to say a word about the relationship between mental models and analogies.

Gentner argues that "The models used in science belong to a large class of analogies that can be characterized as structure-mappings between complex systems" (1980:5). By "structure-mappings" she means that the "relational structure is preserved, but not the objects" (p. 6). She uses Rutherford's atom-as-solar-system analogy as an example: relations like "attracts" and "is larger than" map between nucleus and sun, and "revolves around" maps between electrons and planets. But attributes of the base objects, like the heat and light of the sun, are left behind.

Thus, a good scientific analogy preserves relations but not objects. Gentner also discusses the role of what she calls "base" and "target." The base in the Rutherford case is the solar system; the target is the atom. In essence, the base serves as a mental model that allows construction of a similar model for the target system. But how does one know what features of the base to map onto the target? Here Gentner seems uncertain. Part of the power of mental models derives from their incompleteness and instability; when building a mental model in a new domain by analogy with one in a familiar domain, one is free to incorporate some features and transform or ignore others.

Clement (1988) emphasized the way in which analogies transform knowledge. He asked ten subjects who were either advanced doctoral students or professors in technical areas to consider what would happen if the diameter of the coils of a spring holding a weight was doubled. Would the weight cause the double-diameter spring to stretch farther? Eight of the ten subjects used analogies to solve the problem, and about 60 percent of these analogies involved significant transformations. For example, several subjects considered what would happen if the springs were uncoiled into straight wires. A transformation involves modifying the initial analogy in significant respects. This can lead to new insights: a subject who considered what would happen if the coils in the springs were square discovered the importance of torsion.

Clement's research suggests that analogies can lead to conceptual transformations and breakthroughs. Holton cites as his favorite example of a metaphor or analogy producing a breakthrough the "happiest thought" in Einstein's life: "As with the electric field produced by electro-magnetic induction [1905], the gravitational field has similarly only a relative existence. For if one considers an observer in free fall, e.g., from the roof of a house, there exists for him during his fall no gravitational field—at least in his immediate vicinity" (Einstein, quoted in Holton 1986:235). Einstein's mental model, in this case, was a person in a rising elevator; if this person dropped a ball it would fall in exactly the same way as if the elevator were stationary. To the person inside the rising elevator, the ball drops because of gravity; to an outside observer, the ball drops because the floor rises to meet it. But perhaps there is a way to show the inside observer that he or she is rising. Shine a beam of light through the elevator; if the elevator is rising, the beam will hit the far wall at a slightly lower point. But if the elevator were stationary, light would not be curved merely by gravity. Or would it? This mental model, based on a simple analogy, played an important role in Einstein's prediction that a beam of light would be curved in a gravitational field (Einstein and Infeld 1938). This same prediction, in turn, served as a kind of mental model for Karl Popper's demaraction between science and pseudoscience.

To summarize, the Rutherford example illustrates the way in which a mental model may serve as the "base" for a metaphor or analogy, and the Einstein example illustrates the way in which an analogy can serve as the basis for constructing a mental model.

In the text of chapter 10 (see section 10.2.5) I discuss how Bell used an analogy with the human ear to form his mental model for a speaking telegraph. Similarly, in an 1888 patent caveat, Edison announced that he was "experimenting upon an instrument which does for the eye what the phonograph does for the ear, which is the recording and reproduction of things in motion, and in such a form as to be both cheap, practical and convenient. This apparatus I call a kinetoscope" (Edison, quoted in Carlson and Gorman 1990:396). Edison's analogy between phonograph and kinetoscope was not physiological, like Bell's analogy with nature, but, like Bell's, it gave him a solid basis for a mental model of the new device. Indeed, as Carlson and Gorman show (1990), Edison's kinetoscope was constructed and used in a way very similar to the construction and use of his phonograph; he rejected an alternative based on a different mental model proposed by his assistant Dickson.

25. Recently, heuristics have acquired a bad reputation in some circles, primarily as a result of Tversky and Kahneman (1982), who focus on "cognitive biases that stem from the reliance on judgmental heuristics" (p. 18). A heuristic is a kind of shortcut; therefore it is not completely logical and may lead to errors. Much of the experimental work cited in earlier chapters focused on the heuristic value of confirmation and disconfirmation. Confirmation, if relied on exclusively, can function as a kind of bias, but so can disconfirmation, which can kill a promising idea. The successful application of both of these heuristics depends on the relationship between the subject's mental representation or model and the target rule or law. So one cannot evaluate whether these and other heuristics lead to biases without considering the role of mental models. As Nisbett, Krantz, Jepson, and Fong (1982:447) note, "The nature of one's model for events is of critical importance for the selection of the inferential tools to be used, including various heuristics."

26. Miller (1989) discusses the debate in quantum physics over the role of "visual metaphors," similar to what we would call mental models, in theories. Schrodinger formulated his wave mechanics partly because he felt "repelled by the lack of visualizability" of Heisenberg's approach; Heisenberg, in turn regarded what "Schrodinger writes on visualizability as trash" (Miller 1989:332). Heisenberg later wrote approvingly of Feynman's diagrams, which demonstrates that Heisenberg was only opposed to certain kinds of visualization, linked closely to everyday perception or "customary intuition."

Feynman's theory was formulated in terms of diagrams, and was shown by Dyson to be equivalent to an alternative formulation by Schwinger that had no imaginal content. The relationship between visual theoretical representations, such as Feynman's diagrams, and mental models is a topic for future research.

27. One of the difficulties I faced writing this last chapter is that the telephone research is nearly as complex as the experiments it took me the first six chapters to describe. I have therefore had to condense the case, leaving out most of the details. Those interested in a more thorough comparison of Bell and Gray should look at Gorman, Mehalik, Carlson, and Oblon, in press.

28. In his deposition in *The Telephone Suits,* Part II (1880:142–143), Gray suggests why he dismissed the magneto design: "I thought it would be impossible to make a practical working speaking telephone on the principle shown by Professor Bell, to wit: generating electric currents with the power of the voice, as it seemed to me then that the vibrations were so slight in amplitude and the inductor necessarily so light that the currents thus generated would be too feeble for practical purposes."

29. On October 29, 1877, Gray tried to obtain a patent that included one of his

receiver mechanical representations in the transmitter slot, in lieu of the liquid transmitter. The resulting apparatus was similar in principle to Bell's Centennial telephone, in that it used electromagnetic induction for both the transmitter and the receiver instead of variable resistance for the transmitter and induction for the receiver. This application demonstrates that, while Gray on several occasions accused Bell of stealing his ideas, he was not above borrowing an idea from Bell once in a while. The idea in this case was Bell's mental model for a telephone based on electromagnetic induction, not variable resistance. Gray's patent application was rejected. (See "Gray's Concave Diaphragm Receiver Speaking Telephone Application" in *The Telephone Suits*, Part II, 1880).

30. Peter Slezak (1990), in the paper after mine, argued that these new sociologists were overly influenced by Skinner and other behaviorists. Whether Slezak is right about the affinity between the sociologically inclined philosopher David Bloor and the behaviorists, it is clear that the anti-mentalist position in psychology ran into severe limitations; I suspect it will prove a mistake in science and technology studies as well.

31. On February 16, 1991, Mathey M. Mehalik, W. Bernard Carlson, and I presented a paper to the American Association for the Advancement of Science in which we displayed the current versions of our Bell and Gray maps, and compared them. These maps trace the development of each inventor's mental models and are hierarchically arranged, so that submaps going into further detail are associated with each slot. The maps are stored on a Macintosh computer, allowing us to revise them constantly and add new information as we obtain it. We hope to catalogue and organize virtually every one of Bell's, Gray's, and Edison's telephonic experiments in this way. The History and Philosophy of Science program of the National Science Foundation has generously supported this project (grants No. DIR-8722002 and No. DIR-90-12311). A full account of this research will be provided in a subsequent book; those interested in further information should write to me c/o Division of Humanities, Thornton Hall, University of Virginia, Charlottesville, VA 22903.

Bibliography

Anderson, J. (1983). *The Architecture of Cognition.* Cambridge: Harvard University Press.
Asch, S. E. (1956). Studies of independence and conformity. A minority of one against a unanimous majority. *Psychological Monographs, 70* (9, Whole No. 16).
Baars, B. (1986). *The Cognitive Revolution in Psychology.* New York: Guilford Press.
Banaji, M. R., & Crowder, R. G. (1989). The bankruptcy of everyday memory. *American Psychologist, 44,* 1185–1193.
Barker, P. (1989). The reflexivity problem in the psychology of science. In Gholson, B., Shadish, W. R., Neimeyer, R. A., & Houts, A. C. (Eds.), *Psychology of Science: Contributions to Metascience.* Cambridge: Cambridge University Press, 92–114.
Bazerman, C. (1988). *Shaping Written Knowledge: The Genre and Activity of the Experimental Article in Science.* Madison: The University of Wisconsin Press.
Bell, A. G. (1876a). *The Multiple Telegraph.* Boston: Franklin Press.
——. (1876b). Experiments made by A. Graham Bell, Vol. I, Box 238, Bell Family Papers, Library of Congress.
——. (1908). *The Bell Telephone: Deposition of Alexander Graham Bell.* Boston: American Bell Telephone Co.
Berkowitz, L., & Donnerstein, E. (1982). External validity is more than skin deep: Some answers to criticisms of laboratory experiments. *American Psychologist, 37,* 245–257.
Bernstein, J. (1978). *Experiencing Science.* New York: E. P. Dutton.
Beveridge, W. I. B. (1957). *The Art of Scientific Investigation.* New York: Vintage Books.
Black, M. (1962). *Models and Metaphors.* Ithaca: Cornell University Press.
Blake, C. (1875). The use of the membrana tympani as a phonautograph. *Boston Medical and Surgical Journal, 92,* 121–124.
Boden, M. (1990). *The Creative Mind: Myths & Mechanisms.* New York: Basic Books.
Bradshaw, G. L., Langley, P., & Simon, H. A. (1983). Studying scientific discovery by computer simulation. *Science, 222,* 971–975.
Broadbent, D. E. (1987). Lasting representations and temporary processes. Paper presented in honor of Endel Tulving, University of Toronto, June 11–13.
Brown, H. (1989). Toward a cognitive psychology of what? *Social Epistemology, 3,* 129–138.
Browne, M. W. (1989). Face-off in Italy: What killed the dinosaurs? *The New York Times,* August 15, C1, C11.
Bruce, R. V. (1973). *Bell: Alexander Graham Bell and the Conquest of Solitude.* Boston: Little, Brown.
——. (1976). The American style of invention: Alexander Graham Bell and the strategy of invention. *American Patent Law Association Bulletin,* June, 354–369.
Bruner, J., Goodnow, J., & Austin, G. (1956). *A Study of Thinking.* New York: John Wiley.
Carey, S. (1985). *Conceptual Change in Childhood.* Cambridge: MIT Press.
Carlson, W. B. (1984). Invention, science and business: The professional career of Elihu Thomson, 1870–1900. Ph.D. dissertation, History and Sociology of Science, University of Pennsylvania.
Carlson, W. B., & Gorman, M. E. (1989). Thinking and doing at Menlo Park: Edison's development of the telephone, 1876–1878. In Pretzer, W. (Ed.), *Thomas Edison's Menlo Park Laboratory.* Detroit: Wayne State University Press.
——. (1990). Understanding invention as a cognitive process: The case of Thomas

Edison and early motion pictures, 1888–1891. *Social Studies of Science, 20,* 387–430.

——. (In Press). Using a cognitive framework to understand invention. In Weber, R., & Perkins, D. (Eds.), *The Inventing Mind.* Oxford: Oxford University Press.

Ceci, S. J., & Peters, D. P. (1982). Peer review: A study of reliability. *Change, 14*(6), 44–48.

Chambers, D., & Reisberg, D. (1985). Can mental images be ambiguous? *Journal of Experimental Psychology: Human Perception and Performance, 11,* 317–328.

Chapman, L. J., & Chapman, J. P. (1969). Illusory correlation as an obstacle to the use of valid psychodiagnostic signs. *Journal of Abnormal Psychology, 74,* 271–280.

Chi, M. T. H., Bassok, M., Lewis, M. W., Reimann, P., & Glaser, R. (In Press). Self-explanations: How students study and use examples in learning to solve problems. *Cognitive Science.*

Chi, M. T. H., Feltovich, P. J., & Glaser, R. (1981). Categorization and representation of physics problems by experts and novices. *Cognitive Science, 5,* 121–152.

Christianson, G. E. (1984). *In the Presence of the Creator: Isaac Newton and His Times.* New York: The Free Press.

Churchland, P. M. (1989). *Mind and Science: A Neurocomputational Perspective.* Cambridge: MIT Press.

Cicchetti, D. V., Conn, H. O., & Eron, L. D. (1991). The reliability of peer review for manuscript and grant submissions: A cross-disciplinary investigation. *Behavioral and Brain Sciences, 14,* 119–186.

Clement, J. (1983). A conceptual model discussed by Galileo and used intuitively by physics students. In Gentner, D., & Stevens, A. L. (Eds.), *Mental Models.* Hillsdale, NJ: Lawrence Erlbaum Associates.

——. (1988). Observed methods for generating analogies in scientific problem solving. *Cognitive Science, 12,* 563–586.

Cohen, L. J. (1981). Can human irrationality be experimentally demonstrated? *Behavioral and Brain Sciences, 4,* 317–370.

Collins, H. M. (1985). *Changing Order: Replication and Induction in Scientific Practice.* London: Sage Publications.

——. (1990). *Artificial Experts.* Cambridge: MIT Press.

Collins, H. M., & Pinch, T. J. (1982). *Frames of Meaning: The Social Construction of Extraordinary Science.* London: Routledge & Kegan Paul.

Cosmides, L. (1989). The logic of social exchange: Has natural selection shaped how humans reason? Studies with the Wason selection task. *Cognition, 31,* 187–276.

Crane, D. (1972). *Invisible Colleges: Diffusion of Knowledge in Scientific Communities.* Chicago: University of Chicago Press.

Crouch, T. (1989). *The Bishop's Boys: A Life of Wilbur and Orville Wright.* New York: W. W. Norton.

Crutchfield, R. S. (1955). Conformity and character. *American Psychologist, 10,* 191–198.

Darden, L. (1990). Diagnosing and fixing faults in theories. In Shrager, J., & Langley, P. (Eds.), *Computational Models of Discovery and Theory Formation.* San Mateo, CA: Morgan Kaufmann Publishers, Inc., 319–354.

de Gelder, B. (1989). Granny, the naked emperor, and the second cognitive revolution. In Fuller, S., De Mey, M., Shinn, T., & Woolgar, S. (Eds.), *The Cognitive Turn: Sociological and Psychological Perspectives on Science.* Dordrecht, The Netherlands: Kluwer, 97–118.

de Groot, A. D. (1983). Heuristics, mental programs and intelligence. In

Groner, M., Groner, R., & Bischof, W. F. (Eds.), *Methods of Heuristics.* Hillsdale, NJ: Lawrence Erlbaum Associates, 109–130.

De Kleer, J., & Brown, J. S. (1983). Assumptions and ambiguities in mechanistic mental models. In Gentner, D., & Stevens, A. L. (Eds.), *Mental Models.* Hillsdale, NJ: Lawrence Erlbaum Associates, 155–190.

De Mey, M. (1989). Cognitive paradigms and the psychology of science. In Gholson, B., Shadish, W. R., Neimeyer, R. A., & Houts, A. C. (Eds.), *Psychology of Science: Contributions to Metascience.* Cambridge: Cambridge University Press, 275–295.

Doherty, M. E., & Tweney, R. D. (1988). The role of data and feedback error in inference and prediction. Final report for ARI Contract MDA903-85-K-0193.

Dorner, D. (1983). Heuristics and cognition in complex systems. In Groner, M., Groner, R., & Bischof, W. F. (Eds.), *Methods of Heuristics.* Hillsdale, NJ: Lawrence Erlbaum Associates, 89–108.

Dyer, F. L., Martin, T. C., & Meadowcroft, W. H. (1929). *Edison, His Life and Inventions* (two volumes), New York: Harper's.

Eddington, A. S. (1934/1959). *New Pathways in Science.* Cambridge: Cambridge University Press; reprinted, Ann Arbor: University of Michigan Press.

Edge, D. (1974). Technological metaphor and social control. *New Literary History, 6,* 135–147.

Einhorn, H. J., & Hogarth, R. M. (1978). Confidence in judgment: Persistence of the illusion of validity. *Psychological Review, 85,* 395–416.

Einstein, A., & Infeld, L. (1938). *The Evolution of Physics.* New York: Simon & Schuster.

Epstein, W. M. (1990). Confirmational response bias among social work journals. *Science, Technology and Human Values, 15,* 9–38.

Ericsson, K. A., & Simon, H. A. (1984). *Protocol Analysis: Verbal Reports as Data.* Cambridge: MIT Press.

Evans, J. St. B. (Ed.) (1983). *Thinking and reasoning.* London: Routledge & Kegan Paul.

Faust, D. (1984). *The Limits of Scientific Reasoning.* Minneapolis: University of Minnesota Press.

Ferguson, E. (1977). The mind's eye: Nonverbal thought in technology, *Science, 197,* 827–836.

——. (1980). The nature of technological invention: Play and the intellect. Paper presented to the American Anthropological Association.

Feyerabend, P. (1970). Consolations for the specialist. In Lakatos, I., & Musgrave, A. (Eds.), *Criticism and the Growth of Knowledge.* London: Cambridge University Press.

——. (1975). *Against Method.* Thetford, Norfolk, UK: The Thetford Press.

Finke, R. A. (1989). *Principles of Mental Imagery.* Cambridge, MA: Bradford/MIT Press.

Finn, B. S. (1966). Alexander Graham Bell's experiments with the variable resistance transmitter. *Smithsonian Journal of History, 1* (Winter), 1–16.

Flanagan, O. (1984). *The Science of the Mind.* Cambridge: MIT Press.

Fodor, J. A. (1983). *The Modularity of Mind.* Cambridge: MIT Press.

Freedman, E. (In Press). A computational approach to scientific controversies: The case of latent learning. In R. N. Giere (Ed.), *Cognitive Models of Science. Minnesota Studies in Philosophy of Science.*

Friedel, R., & Israel, P. (1985). *Edison's Electric Light: Biography of an Invention.* New Brunswick: Rutgers University Press.

Fuller, S. (1987). On regulating what is known: A way to social epistemology. *Synthese, 73,* 145–183.

——. (1989). *Philosophy of Science and Its Discontents.* Boulder: Westview Press.

Garcia, J. (1981). Tilting at the paper mills of academe. *American Psychologist, 36*(2), 149–158.
Gardner, H. (1985). *The Mind's New Science: A History of the Cognitive Revolution.* New York: Basic Books.
Gardner, M. (1977). On playing New Eleusis, the game that simulates the search for truth. *Scientific American, 237*(4): 18–25.
Gentner, D. (1980). The structure of analogical models in science. Bolt, Beranek, and Newman Report #4451.
Gentner, D., & Gentner, G. R. (1983). Flowing waters or teeming crowds: Mental models of electricity. In Gentner, D., & Stevens, A. L. (Eds.), *Mental Models.* Hillsdale, NJ: Lawrence Erlbaum Associates, 99–129.
Giere, R. N. (1988). *Explaining Science: A Cognitive Approach.* Chicago: University of Chicago Press.
——. (1989a). Computer discovery and human interests. *Social Studies of Science, 19*(4).
——. (1989b). What does explanatory coherence explain? *Behavioral and Brain Sciences, 12,* 475–476.
Gooding, D. (1985). In Nature's school: Faraday as an experimentalist. In Gooding, D., & James, F. (Eds.), *Faraday Rediscovered: Essays on the Life and Work of Michael Faraday, 1791–1867.* New York: Stockton Press.
——. (1989). "Magnetic curves" and the magnetic field: Experimentation and representation in the history of a theory. In Gooding, D., Pinch, T., & Schaffer, S. (Eds.), *The uses of experiment.* Cambridge: Cambridge University Press.
——. (1990). Mapping experiment as a learning process: How the first electromagnetic motor was invented. *Science, Technology and Human Values, 15*(2), 165–201.
Gooding, D., & Addis, T. (1990). Towards a dynamical representation of experimental procedures. Paper presented at the conference on Rediscovering Skill in Science, Technology and Medicine, Bath, UK, September 14–17.
Goran, M. (1979). *Fact, Fraud and Fantasy: the Occult and Pseudosciences.* New York: A. S. Barnes & Co.
Gorman, Michael E. (1981). Pre-war conformity research in social psychology: The approaches of Floyd H. Allport and Muzafer Sherif. *Journal of the History of the Behavioral Sciences, 17,* 3–14.
——. (1986). How the possibility of error affects falsification on a task that models scientific problem-solving. *British Journal of Psychology, 77,* 85–96.
——. (1987). Will the next Kepler be a computer? *Science & Technology Studies, 5,* 63–65.
——. (1989a). Artificial epistemology? *Social Studies of Science, 19*(2), 374–380.
——. (1989b). Beyond strong programmes. *Social Studies of Science, 19*(4), 643–653.
——. (1989c). Error, falsification and scientific inference: An experimental investigation. *Quarterly Journal of Experimental Psychology, 41A,* 385–412.
——. (1989d). Error and scientific reasoning: An experimental inquiry. In Fuller, S., De Mey, M., Shinn, T., & Woolgar, S. (Eds.), *The Cognitive Turn: Sociological and Psychological Perspectives on Science.* Dordrecht, The Netherlands: Kluwer, 1989, 41–70.
Gorman, Michael E., & Carlson, W. B. (1989). Can experiments be used to study science? *Social Epistemology, 3,* 89–106.
——. (1990). Interpreting invention as a cognitive process: The case of Alexander Graham Bell, Thomas Edison and the telephone. *Science, Technology and Human Values, 15,* 131–164.
Gorman, Michael E., & Gorman, Margaret E. (1984). A comparison of disconfirmatory, confirmatory and a control strategy on Wason's 2-4-6 task. *Quarterly Journal of Experimental Psychology, 36A,* 629–648.

Gorman, Michael E., Gorman, Margaret E., Latta, R. M., & Cunningham, G. (1984). How disconfirmatory, confirmatory and combined strategies affect group problem-solving. *British Journal of Psychology, 75,* 65–79.

Gorman, Michael E., Lind, E. A., & Williams, D. C. (1977). The effects of previous success or failure on a majority-minority confrontation. ERIC Document ED 177 257.

Gorman, Michael E., Mehalik, M. M., Carlson, W. B., & Oblon, M. (In Press). Alexander Graham Bell, Elisha Gray and the speaking telegraph: A cognitive comparison. *History of Technology.*

Gorman, Michael E., Stafford, A., & Gorman, Margaret E. (1987). Disconfirmation and dual hypotheses on a more difficult version of Wason's 2-4-6 task. *Quarterly Journal of Experimental Psychology, 39A,* 1–28.

Griggs, R. (1983). The role of problem content in the selection task and THOG problem. In Evans, J. St. B. T. (Ed.), *Thinking and Reasoning: Psychological Approaches.* London: Routledge & Kegan Paul, 16–43.

Griggs, R., & Cox, J. R. (1982). The elusive thematic-materials effect in Wason's selection task. *British Journal of Psychology, 73,* 407–420.

Groner, M., Groner, R., & Bischof, W. F. (1983). Approaches to heuristics: A historical review. In Groner, M., Groner, R., & Bischof, W. F. (Eds.), *Methods of Heuristics.* Hillsdale, NJ: Lawrence Erlbaum Associates, 1–18.

Gruber, H., & Barrett, P. H. (1974). *Darwin on Man.* New York: Dutton.

Hanson, N. R. (1958). *Patterns of Discovery: An Inquiry into the Conceptual Foundations of Science.* New York: Cambridge University Press.

Haugeland, J. (1986). *Artificial Intelligence: The Very Idea.* Cambridge: MIT Press.

Hawkins, G. S., and White, J. B. (1965). *Stonehenge Decoded.* Garden City, NY: Doubleday & Co.

Hesse, M. B. (1966). *Models and Analogies in Science.* South Bend: University of Notre Dame Press.

Hoffman, R. R. (1980). Metaphor in science. In Honeck, R. P., & Hoffman, R. R. (Eds.), *Cognition and Figurative Language.* Hillsdale, NJ: Lawrence Erlbaum Associates, 393–423.

Holland, J. H., Holyoak, K. J., Nisbett, R. E., & Thagard, P. A. (1986). *Induction: Processes of Inference, Learning and Discovery.* Cambridge: MIT Press.

Holmes, F. (1987). Scientific writing and scientific discovery. *Isis, 78,* 220–235.

Holton, G. (1973). *Thematic Origins of Scientific Thought.* Cambridge: Harvard University Press.

——. (1978). *The Scientific Imagination: Case Studies.* Cambridge: Cambridge University Press.

——. (1986). *The Advancement of Science, and Its Burdens.* Cambridge: Cambridge University Press.

Hon, G. (1989). *Towards a typology of experimental errors: An epistemological view. Studies in History and Philosophy of Science, 20*(4), 469–504.

Hounshell, D. A. (1976). Bell and Gray: Contrasts in style, politics and etiquette. *Proceedings of the IEEE, 64,* 1305–1314.

——. (1981). Two paths to the telephone. *Scientific American, 244*(January), 156–163.

Houts, A. C. (1989). Contributions of the psychology of science to metascience: A call for explorers. In Gholson, B., Shadish, W. R., Neimeyer, R. A., & Houts, A. C. (Eds.), *Psychology of Science: Contributions to Metascience.* Cambridge: Cambridge University Press.

Houts, A., & Gholson, B. (1989). Brownian notions: One historicist philosopher's resistance to psychology of science via three truisms and ecological validity. *Social Epistemology, 3,* 139–146.

Hoyt, W. G. (1976). *Lowell and Mars.* Tucson: The University of Arizona Press.

Hughes, T. P. (1977). Edison's Method. In Pickett, W. B. (Ed.), *Technology at the Turning Point.* San Francisco: San Francisco Press.

Hunt, M. (1982). *The Universe Within.* New York: Simon & Schuster.

Hutchins, E. (1983). Understanding Micronesian navigation. In Gentner, D., & Stevens, A. L. (Eds.), *Mental Models.* Hillsdale, NJ: Lawrence Erlbaum Associates, 191–225.

Insko, C. A., Thibaut, J. W., Moehle, D., Wilson, M., Diamond, W. D., Gilmore, R., Solomon, M. R., & Lipsitz, A. (1980). Social evolution and the emergence of leadership. *Journal of Personality and Social Psychology, 39,* 431–448.

Jacobs, R. C., & Campbell, D. T. (1961). The perpetuation of an arbitrary tradition through several generations of a laboratory microculture. *Journal of Abnormal and Social Psychology, 83,* 649–658.

Jastrow, R. (1983). The dinosaur massacre: A double-barreled mystery. *Science Digest, 91,* 51–53, 109.

Jenkins. R. V. (1984). Elements of style: Continuities in Edison's thinking. *Annals of the New York Academy of Sciences, 424,* 149–162.

Johnson-Laird, P. N. (1983). *Mental Models.* Cambridge: Harvard University Press.

——. (1988). *The Computer and the Mind.* Cambridge: Harvard University Press.

Johnson-Laird, P. N., & Wason, P. C. (1970). Insight into a logical relation. *Quarterly Journal of Experimental Psychology, 22,* 49–61.

——. (1977). A theoretical analysis of insight into a reasoning task. In Johnson-Laird, P. N., & Wason, P. C. (Eds.), *Thinking: Readings in Cognitive Science.* Cambridge: Cambridge University Press.

Josephson, M. (1959). *Edison: A Biography.* New York: McGraw-Hill.

Judson, H. F. (1984). Century of the sciences. *Science 84,5*(9), 41–43.

Karp, P. D. (1990). Hypothesis formation as design. In Shrager, J., & Langley, P. (Eds.), *Computational Models of Discovery and Theory Formation.* San Mateo, CA: Morgan Kaufmann Publishers, Inc., 275–318.

Keller, E. F. (1983). *A Feeling for the Organism: The Life and Work of Barbara McClintock.* New York: W. H. Freeman & Co.

Kern, L. (1982). The effect of data error in inducing confirmatory inference strategies in scientific hypothesis testing. Unpublished Ph.D. dissertation, Ohio State University.

Kern, L. H., Mirels, H. L., & Hinshaw, V. G. (1983). Scientists' understanding of propositional logic: An experimental investigation. *Social Studies of Science, 13,* 131–146.

Kingsbury, J. E. (1915). *The Telephone and Telephone Exchanges: Their Invention and Development.* New York: Longmans, Green. Reprinted Arno, 1972.

Klahr, D., & Dunbar, K. (1988). Dual space search during scientific reasoning. *Cognitive Science, 12,* 1–48.

Klahr, D., Dunbar, K., & Fay, A. L. (1990). Designing good experiments to test bad hypotheses. In Shrager, J., & Langley, P. (Eds.), *Computational Models of Discovery and Theory Formation.* San Mateo, CA: Morgan Kaufmann Publishers, Inc., 355–402.

Klayman, J. (1988). Cue discovery in probabilistic environments: Uncertainty and experimentation. *Learning, Memory, and Cognition, 14,* 317–330.

Klayman, J., & Ha, Y.-W. (1987). Confirmation, disconfirmation and information in hypothesis testing. *Psychological Review, 94,* 211–228.

Klotz, I. M. (1980). The N-ray affair. *Scientific American, 242,* 168–175.

Koestler, A. (1963). *The Sleepwalkers.* New York: Grosset & Dunlap.

Kosslyn, S. M. (1980). *Image and Mind.* Cambridge: Harvard University Press.

Kuhn, T. S. (1957). *The Copernican Revolution.* Cambridge: Harvard University Press.

——. (1962). *The Structure of Scientific Revolutions.* Chicago: University of Chicago Press.

——. (1970). Logic of discovery or psychology of research? Reflections on my critics. In Lakatos, I., & Musgrave, A. (Eds.), *Criticism and the Growth of Knowledge.* London: Cambridge University Press.

Kulkarni, D., & Simon, H. A. (1988). The processes of scientific discovery: The strategies of experimentation. *Cognitive Science, 12,* 139–175.

Kurtz, P. (Ed.) (1985). *A Skeptic's Handbook of Parapsychology.* Buffalo: Prometheus Books.

Lakatos, I. (1978). *The Methodology of Scientific Research Programmes.* Cambridge: Cambridge University Press.

Langley, P., Simon, H. A., Bradshaw, G. L., & Zytkow, J. M. (1987). *Scientific Discovery: Computational Explorations of the Creative Processes.* Cambridge: MIT Press.

Larkin, J. (1983). The role of problem representation in physics. In Gentner, D., & Stevens, A. L. (Eds.), *Mental Models.* Hillsdale, NJ: Lawrence Erlbaum Associates, 75–98.

Larkin, J. H., McDermott, J., Simon, D. P., & Simon, H. A. (1980). Expert and novice performance in solving physics problems. *Science, 208,* 1335–1342.

Latour, B. (1987). *Science in Action.* Cambridge: Harvard University Press.

Latour, B. & Woolgar, S. (1986). *Laboratory Life: The Construction of Scientific Facts.* Princeton: Princeton University Press.

Laudan, L. (1977). *Progress and Its Problems: Towards a Theory of Scientific Growth.* Berkeley: University of California Press.

Laughlin, P. R. (1988). Collective induction: Group performance, social combination processes, and mutual majority and minority influence. *Journal of Personality and Social Psychology, 54,* 254–267.

Laughlin, P. R., & Futoran, G. C. (1985). Collective induction: Social combination and sequential transition. *Interpersonal Relations and Group Processes, 48,* 608–613.

Leary, D. (1980). The intentions and heritage of Descartes and Locke: Toward a recognition of the moral basis of modern psychology *The Journal of General Psychology, 102,* 283–310.

——. (Ed.) (1990). *Metaphors in the History of Psychology.* New York: Cambridge University Press.

Lenat, D. B. (1983). Toward a theory of heuristics. In Groner, M., Groner, R., & Bischof, W. F. (Eds.), *Methods of Heuristics.* Hillsdale, NJ: Lawrence Erlbaum Associates, 351–404.

Lennox, J. G. (1989). Thought experiments: Public and private. Paper presented to the Society for Literature and Science, September 21, Ann Arbor, MI.

McCloskey, M. (1983). Naive theories of motion. In Gentner, D., & Stevens, A. L. (Eds.), *Mental Models.* Hillsdale, NJ: Lawrence Erlbaum Associates, 299–324.

McGuire, W. J. (1989). A perspectivist approach to the strategic planning of programmatic scientific research. In Gholson, B., Shadish, W. R., Neimeyer, R. A., & Houts, A. C. (Eds.), *Psychology of Science: Contributions to Metascience.* Cambridge: Cambridge University Press.

McKinlay, A., & Potter, J. (1987). Model discourse: Interpretative repertoires in scientist's conference talk. *Social Studies of Science, 17,* 443–463.

Mahoney, M. J. (1976). *Scientist as Subject: the Psychological Imperative.* Cambridge, MA: Ballinger.

——. (1977). Publication prejudices: An experimental study of confirmatory bias in the peer review system. *Cognitive Therapy and Research, 1,* 161–175.

——. (1987). Scientific publication and knowledge politics. *Journal of Social Behavior and Personality, 2,* 165–176.
——. (1989). Participatory epistemology and psychology of science. In Gholson, B., Shadish, W. R., Neimeyer, R. A., & Houts, A. C. (Eds.), *Psychology of Science: Contributions to Metascience.* Cambridge: Cambridge University Press.
Mahoney, M. J., & Kimper, T. P. (1976). From ethics to logic: A survey of scientists. In Mahoney, M. J. (1976). *Scientist as Subject: The Psychological Imperative.* Cambridge, MA: Ballinger, 187–193.
Manier, E. (1987). "External Factors" and "Ideology" in the earliest drafts of Darwin's Theory. *Social Studies of Science, 17,* 581–610.
Marks, L. E. (1989). The languages of psychology. *Newsletter of Division 10, Psychology and the Arts, American Psychological Association, Fall/Winter,* 10–20.
Massaro, D. W. (1989). Multiple book review of *Speech Perception by Eye and Ear: A paradigm for psychological inquiry. Behavioral and Brain Sciences, 12,* 741–794.
Masterman, M. (1970). The nature of a paradigm. In Lakatos, I., & Musgrave, A. (Eds.), *Criticism and the Growth of Knowledge.* London: Cambridge University Press.
Mathews, R. C., Buss, R. R., Chinn, R., & Stanley, W. B. (1988). The role of explicit and implicit learning processes in concept discovery. *Quarterly Journal of Experimental Psychology, 40A,* 135–165.
Messeri, P. (1988). Age differences in the reception of new scientific theory: The case of plate tectonics theory. *Social Studies of Science, 18,* 91–112.
Milgram, S. (1974). *Obedience to Authority.* New York: Harper & Row.
Miller, A. I. (1989). Imagery, metaphor, and physical reality. In Gholson, B., Shadish, W. R., Neimeyer, R. A., & Houts, A. C. (Eds.), *Psychology of Science: Contributions to Metascience.* Cambridge: Cambridge University Press, 326–341.
Minsky, M. (1985). *The Society of Mind.* New York: Simon & Schuster.
Mitroff, I. I. (1974). *The Subjective Side of Science.* Amsterdam: Elsevier.
——. (1981). Scientists and confirmation bias. In Tweney, R. D., Doherty, M. E., & Mynatt, C. R. (Eds.), *On Scientific Thinking.* New York: Columbia University Press, 170–175.
Moscovici, S. (1974). Social influence I: Conformity and social control. In Nemeth, C. (Ed.), *Social Psychology: Classic and Contemporary Integrations.* Chicago: Rand McNally, 179–216.
Moscovici, S., & Nemeth, C. (1974). Social influence II: Minority influence. In Nemeth, C. (Ed.), *Social Psychology: Classic and Contemporary Integrations.* Chicago: Rand McNally, 217–249.
Mulkay, M. (1988). Don Quixote's double: A self-exemplifying text. In Woolgar, S. (Ed.), *Knowledge and Reflexivity.* London: Sage.
Myers, G. (1990). *Writing Biology: Texts in the Social Construction of Scientific Knowledge.* Madison: The University of Wisconsin Press.
Mynatt, C. R., Doherty, M. E., & Tweney, R. D. (1977). Confirmation bias in a simulated research environment: An experimental study of scientific inference. *Quarterly Journal of Experimental Psychology, 29,* 85–95.
——. (1978). Consequences of confirmation and disconfirmation in a simulated research environment. *Quarterly Journal of Experimental Psychology, 30,* 395–406.
Neisser, U. (1967). *Cognitive Psychology.* New York: Appleton-Century Crofts.
Newell, A. (1983). The heuristic of George Polya and its relation to Artificial Intelligence. In Groner, M., Groner, R., & Bischof, W. F. (Eds.), *Methods of Heuristics.* Hillsdale, NJ: Lawrence Erlbaum Associates, 195–244.
Newell, A., & Simon, H. A. (1972). *Human Problem Solving.* Englewood Cliffs, NJ: Prentice-Hall.
Nisbett, R. E., Krantz, D. H., Jepson, C., & Fong, G. T. (1982). Improving induc-

tive inference. In Kahneman, D., Slovic, P., & Tversky, A. (Eds.), *Judgment under Uncertainty: Heuristics and Biases.* Cambridge: Cambridge University Press, 3–22.

Norman, D. A. (1983). Some observations on mental models. In Gentner, D., & Stevens, A. L. (Eds.), *Mental Models.* Hillsdale, NJ: Lawrence Erlbaum Associates, 7–14.

——. (1988) *The Psychology of Everyday Things.* New York: Basic Books.

Orne, M. T. (1962). On the social psychology of the psychological experiment with particular reference to the demand characteristics and their implications. *American Psychologist, 17,* 776–783.

Paivio, A. (1986). *Mental Representations: A Dual Coding Approach.* New York: Oxford University Press.

Perkins, D. N. (1981). *The Mind's Best Work.* Cambridge: Harvard University Press.

Peters, D. P., & Ceci, S. J. (1982). Peer review practices of psychology journals. *The Behavioral and Brain Sciences, 5,* 187–255.

Philip, C. O. (1985). *Robert Fulton: A Biography.* New York: Franklin Watts.

Pinch, T. (1985). Theory testing in science—the case of solar neutrinos: Do crucial experiments test theories or theorists? *Philosophy of the Social Sciences, 15,* 167–187.

Platt, J. R. (1964). Strong inference. *Science, 146,* 347–353.

Popper, K. R. (1959). *The Logic of Scientific Discovery.* London: Hutchinson.

——. (1963). *Conjectures and Refutations.* London: Routledge & Kegan Paul.

——. (1970). Normal science and its dangers. In Lakatos, I., & Musgrave, A. (Eds.), *Criticism and the growth of knowledge.* London: Cambridge University Press.

Potter, J., & Wetherell, M. (1987). *Discourse and Social Psychology: Beyond Attitudes and Behaviour.* London & Beverly Hills: Sage.

Prescott, G. B. (1878). *The Speaking Telephone, Talking Phonograph and Other Novelties.* New York: D. Appleton.

——. (1884). *Bell's Electric Speaking Telephone: Its Invention, Construction, Application, Modification and History.* New York: D. Appleton; reprinted Arno, 1972.

Pylyshyn, Z. (1979). Metaphorical impressions and the top-down research strategy. In Ortony, A. (Ed.), *Metaphor and Thought,* Cambridge: Cambridge University Press.

Qin, Y., & Simon, H. A. (1990). Laboratory replication of scientific discovery processes. *Cognitive Science, 14,* 281–312.

Rajamoney, S. (1990). A computational approach to theory revision. In Shrager, J., & Langley, P. (Eds.), *Computational Models of Discovery and Theory Formation.* San Mateo, CA: Morgan Kaufmann Publishers, Inc., 225–254.

Ray, B. C. (1987). Stonehenge: A new theory. *History of Religions, 26,* 225–278.

Reichenbach, H. (1938). *Experience and Prediction.* Chicago: University of Chicago Press.

Rensberger, B. (1986). Death of the Dinosaurs: A true story? *Discover,* May, 28–35, 107–108.

Rhodes, F. L. (1929). *Beginnings of Telephony.* New York: Harper & Bros.

Rosenwein, R. E. (In Press). The role of dissent in achieving scientific consensus. In Shadish, W. R. (Ed.), *Social Psychology of Science.* New York: Guilford Press.

Rouse, W. B., & Morris, N. M. (1986). On looking into the black box: Prospects and limits in the search for mental models. *Psychological Bulletin, 100,* 349–363.

Rudwick, M. J. S. (1985). *The Great Devonian Controversy.* Chicago: University of Chicago Press.

Schanck, R., & Abelson, R. (1977). *Scripts, Plans, Goals and Understanding: An Inquiry into Human Knowledge Structures.* Hillsdale, NJ: Lawrence Erlbaum Associates.

Shadish, W. R. (1989). The perception and evaluation of quality in science. In Gholson, B., Shadish, W. R., Neimeyer, R. A., & Houts, A. C. (Eds.), *Psychology of Science: Contributions to Metascience.* Cambridge: Cambridge University Press.
——. (Ed.). (In Press). *Social Psychology of Science.* New York: Guilford Press.
Shapin, S. (1982). History of science and its sociological reconstruction. *History of Science, 20,* 157–211.
Sherif, M. (1936). *The Psychology of Social Norms.* New York: Harper.
Sherif, M., Harvey, O. J., White, B. J., Hood, W. R., & Sherif, C. W. (1961). *Intergroup Conflict and Cooperation: The Robers Cave Experiment.* Norman: University of Oklahoma Book Exchange.
Shrager, J. (1990). Commonsense perception and the psychology of theory formation. In Shrager, J., & Langley, P. (Eds.), *Computational Models of Scientific Discovery and Theory Formation.* San Mateo, CA: Morgan Kaufmann Publishers, Inc., 437–470.
Shrager, J., & Langley, P. (1990). *Computational Models of Scientific Discovery and Theory Formation.* San Mateo, CA: Morgan Kaufmann Publishers, Inc.
Sieber, J. E. (1989). On studying the powerful (or fearing to do so): A vital role for IRBs. *IRB: A Review of Human Subjects Research. 11,* 1–6.
Siegel, H. (1980). Justification, discovery and the naturalization of epistemology. *Philosophy of Science, 47,* 297–321.
Simon, H. A. (1988). Knowledge and search in expert systems: Applications to scientific discovery. Paper presented at Michigan Technological University, November 20.
——. (1989). ECHO and STAHL: On the theory of combustion. *Behavioral and Brain Sciences, 12,* 487.
Simon, H. A., Langley, P. W., & Bradshaw, G. (1981). Scientific discovery as problem solving. *Synthese, 47,* 1–27.
Singley, M. R., & Anderson, J. R. (1989). *The Transfer of Cognitive Skill.* Cambridge: Harvard University Press.
Slezak, P. (1989). Scientific discovery by computer as empirical refutation of the Strong Programme. *Social Studies of Science, 19*(4), 563–600.
——. (1990). Behaviorism and the strong programme. Paper presented to the Society for the Social Studies of Science, Minneapolis, MN, October 19.
Smith, L. D. (1986). *Behaviorism and Logical Positivism.* Stanford: Stanford University Press.
Spencer, N. J., Hartnett, J., & Mahoney, M. J. (1986). Problems with reviews in the standard editorial practice. *Journal of Social Behavior and Personality, 1,* 21–36.
Steiner, I. D. (1972). *Group Process and Productivity.* New York: Academic Press.
Stich, S. P. (1985). Could man be an irrational animal? *Synthese, 64,* 115–135.
Suchman, L. A. (1987). *Plans and Situated Actions: The Problem of Human-Machine Interaction.* Cambridge: Cambridge University Press.
Taubes, G. (1986). The game of the name is fame, but is it science? *Discover,* December, 28–52.
Taylor, L. (Unpublished Manuscript). *The untold story of the telephone.* Elisha Gray Papers, National Museum of American History, Box 6, Folder 6.
The Telephone Suits: Bell Telephone Company Et Al. v. Peter A. Dowd. Part II: Exhibits of Complainants and Defendant. (1880). Boston: Alfred Judge & Son, Law Printers.
Thagard, P. (1988). *Computational Philosophy of Science.* Cambridge: MIT Press.
——. (1989). Explanatory coherence. *Behavioral and Brain Sciences, 12,* 435–467.
Thagard, P., Holyoak, K. J., Nelson, G., and Gochfeld, D. (1990). Analog retrieval by constraint satisfaction. *Artificial Intelligence, 46,* 259–310.

Thagard, P., & Nowak, G. (1990). The conceptual structure of the geological revolution. In Shrager, J., & Langley, P. (Eds.), *Computational Models of Scientific Discovery and Theory Formation.* San Mateo, CA: Morgan Kaufmann Publishers, Inc, 27–72.

Tukey, D. D. (1986). A philosophical and empirical analysis of subjects' modes of inquiry on the 2-4-6 task. *Quarterly Journal of Experimental Psychology, 38A,* 5–33.

Tversky, A., & Kahneman, D. (1982). Judgment under uncertainty: Heuristics and biases. In Kahneman, D., Slovic, P., & Tversky, A. (Eds.), *Judgment under Uncertainty: Heuristics and Biases.* Cambridge: Cambridge University Press, 3–22.

Tweney, R. D. (1984). Cognitive psychology and the history of science: a new look at Michael Faraday. In Reppard, H., van Hoorn, W., & Bem, S. (Eds.), *Studies in the History of Psychology and the Social Sciences, 2,* Leiden: Psychologisch Instituut van der Ryks—Universiteit Leiden, 235–246.

——. (1985). Faraday's discovery of induction: A cognitive approach. In Gooding D., & James, F. (Eds.), *Faraday Rediscovered: Essays on the Life and Work of Michael Faraday, 1791–1867.* New York: Stockton Press.

——. (1986). Procedural representation in Michael Faraday's scientific thought. *PSA 1986, 2,* 336–344.

——. (1989). A framework for the cognitive psychology of science. In Gholson, B., Shadish Jr., W. R., Neimeyer, R. A., & Houts, A. C. (Eds.), *Psychology of Science.* Cambridge: Cambridge University Press.

Tweney, R. D., Doherty, M. E. (1983). Rationality and the psychology of inference. *Synthese, 57,* 139–161.

Tweney, R. D., Doherty, M. E., & Mynatt, C. R. (Eds.). (1981). *On Scientific Thinking.* New York: Columbia University Press.

Tweney, R. D., Doherty, M. E., Worner, W. J., Pliske, D. B., Mynatt, C. R., Gross, K. A., & Arkkelin, D. L. (1980). Strategies of rule discovery on an inference task. *Quarterly Journal of Experimental Psychology, 32,* 109–123.

Tweney, R. D., & Yachanin, S. A. (1985). Can scientists assess conditional inferences? *Social Studies of Science, 15,* 155–173.

Walker, B. J., Doherty, M. E., & Tweney, R. D. (1987). The effects of feedback error on hypothesis testing. Paper presented at the annual meeting of the Midwestern Psychological Association, Chicago, IL.

Waltz, D. L. (1987). Applications of the Connection Machine. *Computer,* January, 85–97.

——. (1988). Artificial intelligence. *Daedalus, 117,* 191–212.

Wason, P. C. (1960). On the failure to eliminate hypotheses in a conceptual task. *Quarterly Journal of Experimental Psychology, 12,* 129–140.

——. (1962). Reply to Wetherick. *Quarterly Journal of Experimental Psychology, 14:* 250.

——. (1966). Reasoning. In Foss, B. (Ed.), *New Horizons in Psychology: I.* Baltimore: Penguin.

——. (1977). On the failure to eliminate hypotheses . . . A second look. In Johnson-Laird, P. N., & Wason, P. C. (Eds.), *Thinking: Readings in Cognitive Science.* Cambridge: Cambridge University Press.

——. (1983). Realism and rationality in the selection task. In Evans, J. St. B. T. (Ed.). *Thinking and Reasoning: Psychological Approaches.* London: Routledge & Kegan Paul.

Wason, P. C., & Green, D. W. (1984). Reasoning and mental representation. *Quarterly Journal of Experimental Psychology, 36A,* 597–610.

Wason, P. C., & Johnson-Laird, P. N. (1970). A conflict between selecting and evaluating information in an inferential task. *British Journal of Psychology, 61,* 509–515.

Watson, T. A. (1913). The birth and babyhood of the telephone. Address delivered before the 3rd Annual Convention of the Telephone Pioneers of America, Chicago, October 17. Reprinted by the American Telephone & Telegraph Co., 1977.

Weber, R. J., & Dixon, S. (1989). Invention and gain analysis. *Cognitive Psychology, 21,* 1–21.

Weber, R. J., & Perkins, D. N. (1989). How to invent artifacts and ideas. *New Ideas in Psychology, 7,* 49–72.

Westrum, R., & Wilcox, H. A. (1989). Sidewinder. *Invention & Technology,* Fall issue, 57–63.

Wetherick, N. E. (1962). Eliminative and enumerative behaviour in a conceptual task. *Quarterly Journal of Experimental Psychology, 14,* 246–249.

Wilensky, R. (1983). *Planning and Understanding: A Computational Approach to Human Reasoning.* Reading, MA: Addison-Wesley.

Williams, L. P. (1965). *Michael Faraday: A Biography.* New York: Basic Books.

Wiser, M., & Carey, S. (1983). When heat and temperature were one. In Gentner, D., & Stevens, A. L. (Eds.), *Mental Models.* Hillsdale, NJ: Lawrence Erlbaum Associates, 267–297.

Woolgar, S. (1987). Reconstructing man and machine: A note on sociological critiques of cognitivism. In Bjiker, W. E., Hughes, T., & Pinch, T. (Eds.), *The Social Construction of Technological Systems.* Cambridge: MIT Press.

——. (Ed.). (1988). *Knowledge and Reflexivity.* London: Sage.

——. (1989). A coffeehouse conversation on the possibility of mechanising discovery and its sociological analysis, with some thoughts on 'decisive refutations', 'adequate rebuttals' and the prospects for transcending the kinds of debate about the sociology of science of which this is an example. *Social Studies of Science, 19*(4), 658–668.

Zucker, L. G. (1977). The role of institutionalization in cultural persistence. *American Sociological Review, 42,* 726–743.

Index

MICHAEL E. GORMAN is Associate Professor in the Humanities Division of the School of Engineering and Applied Sciences at the University of Virginia.